Bo Zhang

Soft Magnetic Composites in Novel Designs of Electrical Traction Machines

Soft Magnetic Composites in Novel Designs of Electrical Traction Machines

by

Bo Zhang

Scientific Publishing

Dissertation, Karlsruher Institut für Technologie (KIT)
Fakultät für Elektrotechnik und Informationstechnik

Tag der mündlichen Prüfung: 17. November 2016
Referenten: Prof. Dr.-Ing. Martin Doppelbauer
Prof. Dr.-Ing. Nejila Parspour (Universität Stuttgart)

Impressum

KIT Scientific Publishing

Karlsruher Institut für Technologie (KIT)
KIT Scientific Publishing
Straße am Forum 2
D-76131 Karlsruhe

KIT Scientific Publishing is a registered trademark of Karlsruhe Institute of Technology. Reprint using the book cover is not allowed.

www.ksp.kit.edu

Print on Demand 2017 – Gedruckt auf FSC-zertifiziertem Papier

ISBN 978-3-7315-0619-5
DOI 10.5445/KSP/1000064348

Soft Magnetic Composites in Novel Designs of Electrical Traction Machines

Zur Erlangung des akademischen Grades eines

DOKTOR-INGENIEURS

an der Fakultät für
Elektrotechnik und Informationstechnik
des Karlsruher Instituts für Technologie (KIT)
genehmigte

DISSERTATION

von

M.Sc. Bo Zhang
geb. in: Shaanxi, China

Tag der mündlichen Prüfung:	17.11.2016
Hauptreferent:	Prof. Dr.-Ing. Martin Doppelbauer
Korreferent:	Prof. Dr.-Ing. Nejila Parspour (Universität Stuttgart)

Contents

Vorwort

Die vorliegende Dissertation entstand während meiner Untersuchung und Arbeit als wissenschaftlicher Mitarbeiter im Lehrstuhl Hybridelektrische Fahrzeuge (HEV) des Elektrotechnischen Instituts (ETI) am Karlsruher Institut für Technologie (KIT) in Zusammenarbeit mit der Daimler AG und PMG Füssen GmbH. Allen die mir während der Promotion geholfen und zum Gelingen dieser Arbeit beigetragen haben, möchte ich an dieser Stelle ganz herzlich danken.

Herrn Prof. Dr.-Ing. Martin Doppelbauer danke ich für das entgegengebrachte Vertrauen und die starke Unterstützung, wodurch ich große Freiheit und Motivation während der Untersuchung in diesem interessanten Themengebiet haben könnte. Die unzähligen fachlichen Diskussionen mit ihm haben maßgeblich zum Gelingen dieser Arbeit beigetragen. Ebenso danke ich Herrn Prof. Dr.-Ing. xx für sein Interesse an der Arbeit und die Übernahme des Korreferats.

Außerdem bedanke ich Herrn Prof. Dr.-Ing. Michael Braun für seine Betreuung und Hilfe. Vom Zuhören seiner Vorlesung als Masterstudent, bis zur Zusammenarbeit mit ihm als wissenschaftlicher Mitarbeiter am ETI, habe ich unglaublich viel von ihm gelernt, was für mich lebenslang hilfreich sein wird.

Den industriellen Partnern, insbesondere Herrn Matthias Gregor und Herrn Dr.-Ing. Robert Inderka der Daimler AG sowie Herrn Maximilian Sigl und Herrn Dr.-Ing. Andreas Schoppa der PMG Füssen GmbH danke ich für die großartige Unterstützung bei der Zusammenarbeit.

Des Weiteren bedanke ich mich bei allen Kollegen am ETI, insbesondere Herrn Torsten Epskamp, der mit mir für lange Zeit in einem Büro zusammengearbeitet hat. Die fachliche und persönliche Zusammenarbeit bereitete mir große Freude. Die Kollegen haben mir und meiner Frau sogar Squash und Skifahren bei-gebracht, was uns viel große Freude gebracht haben.

Beim Prototypaufbau der Axialflussmaschine haben die Kollegen der Werkstatt mich stark unterstützt. Hervorzuheben sind an der Stelle Herr Rouven Burger und Herr Bernhard Saschin. Ihre Erfahrung und Vorschläge waren stets äußerst konstruktiv.

Ein großer Dank geht auch an die Studenten, die im Rahmen ihrer studentischen Arbeiten Teile zum Fortschritt meiner Arbeit beigetragen haben. Mit ihnen habe ich unzählige fachliche Diskussionen geführt, die einige Aspekte dieser Dissertation erst ermöglicht haben.

Ich wollte auch meiner Familie und insbesondere meiner Frau ganz herzlich danken, die mir während des Studiums und der Promotion großen Rückhalt gegeben haben. Meinen Eltern und meiner Schwester danke ich ganz besonders für alles, was sie in der vergangen Zeit für mich getan haben.

Allen weiteren Freunden, Bekannten und Verwandten gilt ebenfalls mein herzlicher Dank. Obwohl viele davon Deutschland schon verlaßen haben, die Zeit mit euch ist immer die schönste Erinnerung von mir.

Bo Zhang

Karlsruhe, 17.11.2016

序言

您正在翻阅的博士论文源于我在卡尔斯鲁厄理工学院(Karlsruher Institute of Technology)电气工程所(Institute of Electrical Engineering)博士学习期间所完成的研究工作。我想借此机会，对于所有给过我建议和帮助的人表达我衷心的感谢和内心最美好的祝愿。

首先我想感谢我的博士生导师Martin Doppelbauer教授。正是因为他在博士期间对我大力的帮助和充分的信任，使我在科研中有了极大的自由空间。他适时的指导，却又总能让我在迷惘时能找到新的研究灵感和方向。他的每一次鼓励和批评，都能让我受益匪浅，获益良多，也最终促成了我博士论文的顺利完成。

其次，我要感谢Michael Braun教授。从第一次作为研究生听他所教授的电机控制课程，到后来博士期间和他逐渐熟识，我看到了德国最优秀工程师的对事业孜孜不倦的执着和坚持，以及科研中严谨，求实，一丝不苟的态度。相信在未来的工作中，他也是我时时学习的榜样。

因为我的研究项目是基于和Daimler AG以及PMG Füssen GmbH公司的合作，我还要感谢我的公司导师Matthis Gregor，Robert Inderka博士，Maximilian Sigl以及Andreas Schoppa博士在研究工作中所提供的无私建议和帮助。

此外我还要感谢和我一起在研究所工作的所有同事们，尤其是和我在一间办公室工作的Torsten Epskamp。他们不但和我有过许多次深入的学术讨论，也在业余时间和我一起度过了很多美好，难忘的时光。值得一提的是，同事们甚至还教会了我滑雪和壁球。因为他们，我更好，更快的适应了在德国的工作和生活。

在电机样机制造阶段，研究所工作室的同事们也对我提供了有力的支持和帮助，尤其是Bernhard Saschin。他多年的机械加工制造经验，对样机的设计和制造绝对是必不可少。

我还要感谢所有我指导过的学生们，他们的学士和硕士论文是我博士研究的扎实基础。因为他们的巨大付出，我才能顺利完成研究项目。

我还要感谢我的家人，尤其是我的妻子。这些年，她对家庭巨大的默默付出，使我能够心无旁骛的完成研究。没有她对我长期默默的巨大支持，我不可能如此顺利的完成博士学业。

最后，我还要感谢所有亲人和朋友。虽然很多人已经离开卡鲁，但同你们共度的时光将是我深藏在心底里最美好的回忆。

张博

卡尔斯鲁厄, 17.11.2016

Zusammenfassung

Heutzutage werden hauptsächlich laminierte Elektrobleche eingesetzt, um die weichmagnetischen Bauteile einer elektrischen Maschine zu fertigen. Im Vergleich dazu bestehen weichmagnetische Pulververbundwerkstoffe (Englisch: soft magnetic composites – SMC) aus vielen kleinen voneinander isolierten Eisenpartikeln.

Dank der isotropen Struktur besitzen die Pulververbundwerkstoffe viele besondere Eigenschaften. Der spezifische Widerstand von SMC ist sehr hoch, was zu niedrigen Wirbelstromverlusten führt. Pulververbundwerkstoffe zeichnen sich durch thermische und magnetische Isotropie aus. Viele neuartige elektrische Maschinen mit einer dreidimensionalen Flussführung werden durch die isotropen Eigenschaften erst ermöglicht. Außerdem wird das Pulver direkt in die gewünschte Form gepresst und anschließend wärmebehandelt. Dadurch können Bauteile mit komplizierter Geometrie einfach und effizient gefertigt werden.

In der vorliegenden Arbeit wurde die Anwendung von SMC in neuen Konzepten elektrischer Maschinen untersucht und bewertet, speziell im Hinblick auf den Einsatz in Kraftfahrzeugen als Antriebsmaschine. Diese Dissertation besteht hauptsächlich aus den folgenden Arbeiten:

- Messung der magnetischen Eigenschaften verschiedener Pulververbundwerkstoffe
- Entwicklung und Validierung eines Modells zur Berechnung der Eisenverluste in Pulververbundwerkstoffen
- Entwicklung einer Simulationskette auf Basis der 3D FEM zur Analyse verschiedener neuartiger Maschinen

- Elektromagnetische Auslegung einer Axialflussmaschine und einer Transversalflussmaschine
- Mechanische Konstruktion und Analyse der ausgelegten Axialflussmaschine mit einem Stator und zwei äußeren Rotoren
- Prototypaufbau und Messung der Axialflussmaschine zur Validierung der Simulationsergebnisse

Auf Basis geschaffter Arbeiten wird gezeigt, dass für die vorgegebenen Randbedingungen neuartige elektrische Maschinen z.B. Axialflussmaschine vergleichbare Eigenschaften wie traditionelle Radialflussmaschine erreichen können. Die Axialflussmaschine ist eine konkurrenzfähige Topologie zur Radialflussmaschine, wenn eine kurze Axiallänge, hohe Drehmoment- und Leistungs-dichte gleichzeitig gefordert werden. Obwohl die Transversalflussmaschine wegen möglicher hoher Polpaarzahlen eine hohe Drehmomentdichte erreichen kann, steht dem Einsatz als Antriebsmaschine ein relativ niedriger Leistungsfaktor aufgrund der hohen Streuflüsse entgegen.

Glossary

Symbols

A	$\mathrm{Wb\,m^{-1}}$	magnetic vector potential
B	T	magnetic flux density
B_{r}	T	Remanence
H	$\mathrm{A\,m^{-1}}$	magnetic field strength
H_{c}	$\mathrm{A\,m^{-1}}$	Coercivity
J	T	magnetic Polarisation
J	$\mathrm{A\,mm^{-2}}$	electrical current density
J_{s}	T	saturation polarisation
T	N m	Torque
P_{el}	W	iron loss
U	V	electrical voltage
d	mm	thickness
f	Hz	electrical frequency
f		function, in general
l_{mag}	mm	length of magnetic path
m		number of phases
m	kg	mass
n	$\mathrm{min^{-1}}$	rotational speed
p		number of pole-pairs
t	s	time

Ω	$\mathrm{rad\,s^{-1}}$	mechanical angular velocity
Φ	Wb	magnetic flux
Ψ	Wb	linked magnetic flux
η		efficiency
μ	$\mathrm{H\,m^{-1}}$	permeability
μ_{r}		relative permeability
μ_0	$\mathrm{H\,m^{-1}}$	vacuum permeability
ω	$\mathrm{rad\,s^{-1}}$	electrical frequency
ρ_{el}	$\Omega\,\mathrm{m}$	electrical resistivity
ρ_{m}	kg	density
σ	$\mathrm{S\,m^{-1}}$	electrical conductivity

Abbreviations

2D	two dimensional
3D	three dimensional
AFM	axial flux machine
AMM	amorphous magnetic metal
ANOVA	analysis of variance
ASM	asynchronous machine
CFD	computational fluid dynamic
DLCW	double layer concentrated winding
DoE	design of experiment method
DSP	digital signal processor
EM	electrical machine
EMF	electromotive force
FDW	flux density waveform
FEM	finite element method
FL	fuzzy logic
GA	genetic algorithms
GUI	graphical user interface
IGBT	insulated gate bipolar transistor

LFM	longitudinal flux machine
LHS	latin hypercube sampling
MMF	magnetomotive force
MOPSO	multi-objective particle swarm optimization
NES	non-grain oriented electrical steel
NRMSE	normalized root mean squared error
NSGA2	non dominated sorting genetic algorithm 2
OA	orthogonal array
PEEK	polyetheretherketon
PM	permanent magnet
PMAFM	PM excited axial flux machine
PMAFM-IR	PMAFM with internal rotor
PMAFM-IS	PMAFM with internal stator
PMAFM-ISA	PMAFM with internal segmented armature, also named as YASA or SAT
PMRFM	PM excited radial flux machine
PMTFM	PM excited transverse flux machine
PMTFM-CP	PMTFM with claw poles
POM	polyoxymethylene
RBF	radial basis function
RFM	radial flux machine
RTFM	reluctance transverse flux machine
SAT	segmented armature torus machine
SMC	soft magnetic composites
SRM	switched reluctance machine
TFM	transverse flux machine
YASA	yokeless and segmented axial flux machine

1 Introduction

1.1 Soft magnetic composites

In recent years, a lot of researches have been carried out in the fields of electrical machines, power electronics and batteries because of rising interest in electric mobility motivated by the climate change caused by emission of CO_2, finiteness of fossil fuels and increasing demands on the protection of the environment [1, 2]. Different from traditional industrial applications, the total available energy and space in electric mobility are severely limited. For this reason, there are some special requirements on the used electrical machines such as high efficiency, high torque and power density, wide speed range, considerable maintainability and reliability.

Owing to the rapid development of material science, the performance of electrical machine has been continuously improved in the past decades. For instance, efficiency and power density have been enhanced through the application of rare-earth permanent magnets (PMs), which are well known for their high energy density. Nowadays, non-grain oriented electrical steel (NES) is widely utilized to manufacture magnetic cores of electrical machine. It is an iron alloy with typically $0 - 3\%$ silicon as additive.

For the application in radial flux machine, the coated laminations are usually punched and stacked together to avoid excessive eddy current losses caused by flux passing perpendicularly through the laminations. Instead of silicon, there are also cobalt-iron and nickel-iron alloys that are distinguished by high saturation flux density and high permeability at the cost of high price [3–5].

Because of the outstanding mechanical properties, high saturation flux density and permeability, manufacturing of electrical machines based on electrical steel laminations has been well investigated and established. However, the eddy current losses increase significantly at high frequencies, which is one of the reasons that limits the attainable speed of an electrical machine. An effective method to reduce the eddy current is to utilize thinner electrical steel sheets, which however leads to a much higher price [6]. In addition, the magnetic flux can only be two-dimensionally conducted within a plane due to insulation among the lamination sheets. It is the reason why the basic construction of electrical machine has remained unchanged in the past time.

As a potential competitor to NES, the newly developed amorphous magnetic metal (AMM) is featured by its disordered atomic structure and much smaller thickness. The AMM is manufactured by rapidly freezing molten iron at high temperature to avoid the formation of equilibrium crystalline atomic structure [7]. Due to its special structure, the AMM can offer obviously improved mechanical and electromagnetic properties such as high permeability and extremely low iron losses. Thus, AMM has attracted lots of attentions from researchers and many prototypes have been developed. From the perspective of magnetics, the challenges with AMM are low saturation flux density and stacking factor which can limit the maximal magnetic loading significantly. Besides, the AMM becomes extremely brittle after the thermal treatment, which is difficult to be manufactured with conventional cutting or punching.

Another important impulse promoting the development of new topologies of the electrical machine comes from the soft magnetic composites (SMC), whose mechanical and magnetic properties have been significantly improved in the last decades. Different from the laminated steel, the SMC comprise a large number of fine iron particles coated with the insulation material, as illustrated in Fig. 1.1.

Due to the unique structure, the SMC have many unique characteristics that can improve the performance of traditional electrical machine and development of novel designs with more complex magnetic circuits.

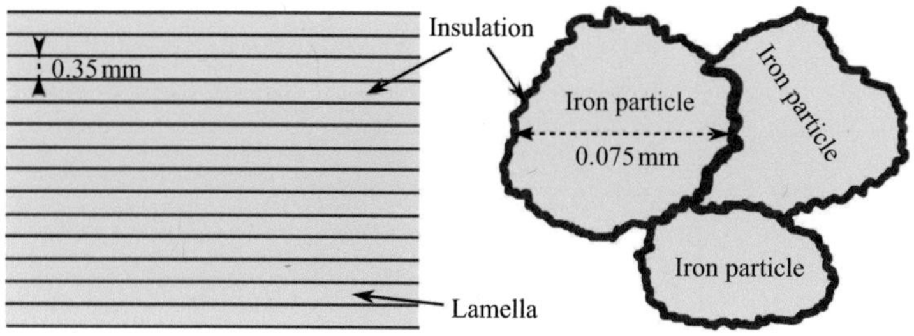

Figure 1.1. Schematic structure of NES and SMC

Isotropy

First of all, SMC are isotropic implying that the physical properties are independent of the orientations. This characteristic is valid both for thermal and magnetic properties. Therefore, the heat derived from losses can be effectively dissipated. The magnetic isotropy is potentially a more important property. Due to the magnetic isotropy, electrical machines with complex three-dimensional (3D) magnetic pattern can be realized, which are difficult or even impossible to be manufactured with traditional laminated steel.

High electrical resistance

The electrical resistance of SMC is much higher than that of NES because of thin insulation layer around each iron particle, which can effectively reduce eddy current losses. This indicates that the SMC-based electrical machines are more advantageous at high operating frequency where eddy current losses become the dominant component of iron losses. For the inverter-fed electrical machine, especially the reluctance machine, the expected current waveform is far away from sinusoidal. As a result, there are generally more harmonics contained in the magnetic field, which makes the application of SMC more considerable. The typical applicable regions of SMC, ferrite and laminated electrical steel in terms of saturation flux density and frequency are illustrated in Fig. 1.2.

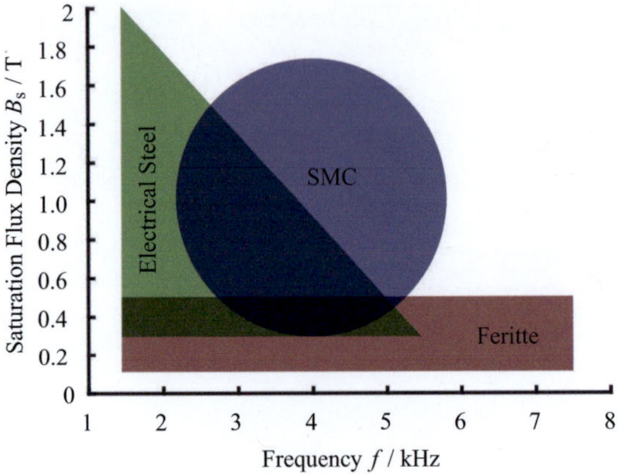

Figure 1.2. Application regions of SMC, electrical steel and ferrite [5, 8, 9]

Production techniques

The manufacturing process of SMC mainly consists of only compaction and heat treatment, which is simple and cost effective in large volumes compared with that of electrical steel. Besides, the soft magnetic components with complex geometry, high surface quality and tight tolerance can be manufactured with SMC. For instance, some soft magnetic parts made from SMC are illustrated in Fig. 1.3.

Figure 1.3. Some motor components made from SMC [10–12]

High material utilization and recycling ability

Furthermore, the waste of raw material during the manufacturing process of SMC is extremely limited. In addition, the SMC cores enable high recycling ability of raw material because copper winding can be easily separated from SMC. In summary, a high material utilization can be achieved, as shown in Fig. 1.4.

Figure 1.4. Recycling of motors with SMC core and lamination core [10]

Considering above advantages, it can be concluded that components with 3D magnetic flux conductivity and complex shape can be manufactured with SMC. Thus, great opportunities have been opened up for the researchers to design novel electrical machines with innovative shape and complex magnetic circuit.

However, some obvious disadvantages preventing the widespread application of SMC, which should be taken into account during the design of the electrical machine. Some of them are listed as follows:

- Low magnetic permeability
- High hysteresis losses
- Inferior mechanical properties
- Limited geometrical size of the available parts

From the viewpoint of magnetics, the relative permeability of insulation layer around the iron particle is close to air. Therefore, the unsaturated magnetic permeability and saturation flux density of SMC are much lower than that of electrical steel. For the same reason, the SMC are not suitable for the electrical machine requiring high magnetizing current such as reluctance machine. On the contrary, the applications in machines with large effective air gaps such as PM excited synchronous machines are more suitable for the application of SMC.

In addition, the SMC is inferior to NES in terms of mechanical properties, which makes the mechanical construction a great challenge. Furthermore, the SMC have much higher hysteresis losses which are the dominant component of iron losses in low-frequency domain. At last but not least, a high pressure is necessary during the compaction of the mixed powder. Thus, a larger press machine is needed if dimensions of the demanded SMC parts are large. As a result, the available size of components made from SMC is limited.

The above conclusions can be deduced from the following Table 1.1, where typical properties of three commercially available materials including NES, AMM and SMC are compared.

		NES ThyssenKrupp M235-35A	**AMM** METGLAS 2605SA1	**SMC** Höganäs 3P Soma-loy700HR
Lamination thickness	[mm]	0.35	0.025	solid
ρ_{el}	[$\mu\Omega$cm]	48	130	15,840
Density	[kg/m^3]	7600	7180	7510
Stacking factor	[%]	95	≥84	-
P_{fe} at 50 Hz, 1 T	[W/kg]	0.95	0.05	4.53
P_{fe} at 50 Hz, 1.5 T	[W/kg]	2.35	0.19	9.05
P_{fe} at 1 kHz, 1 T	[W/kg]	69.8	6.5	137
Saturation flux density	[T]	2.1	1.6	1.9
Maximum μ_r	-	6,800	600,000	760
Tensile strength	[N/mm^2]	532	2000	120 [TRS]
Maximal standard available width	[mm]	1250	213	-
Magnetic flux path	-	2D	2D	3D

Table 1.1. Comparison of the properties of NES, AMM and SMC

1.2 Electrical traction machine

Electrical machines were developed almost 200 years ago. Since the invention of the first practicable rotational electrical machine in 1834 by Moritz Hermann Jacobi, a large number of electrical machines with different configurations were invented. Compared with other applications, there are some special requirements for the electrical traction machine in the hybrid electric vehicle (HEV) and electric vehicle (EV) [13]:

- High compactness, small size and low weight
- High peak efficiency and high driving cycle efficiency
- High torque and power density
- Large speed range with constant maximum torque
- Excellent robustness and maintenance
- Low vibration and acoustic noise level

Nowadays, the radial flux machine (RFM) has been already widely accepted as the traction machines. Among the different kinds of RFM, the permanent magnet (PM) excited RFM is the most frequently used topology due to its high efficiency and high power density [14, 15]. These advantages are mainly resulted from the application of PM with high energy density, which are of great importance considering the high transient torque requirement and the limited space in the HEV. The PMs can be either mounted on the surface of the rotor yoke or embedded in the rotor. The latter type is more favorable due to the existence of reluctance torque in addition to the synchronous torque, which is resulted from the difference between the inductances along the direct axis (d-axis) and lateral axis (q-axis) [16–18].

The application of SMC in almost all kinds of traditional RFM has been investigated, as described in section 2.5. It has been shown that a simple replacement of NES with SMC in RFM usually leads to poorer performance due to the disadvantages of SMC such as low unsaturated permeability and high hysteresis losses.

As a result, more and more novel designs of electrical traction machine have been suggested in the past years to fully utilize the unique properties of SMC. Among the different novel designs, the axial flux machine (AFM) becomes a hot spot of researches due to its short axial length, compactness and high torque density. The AFM is characterized by the main flux flowing axially in the air gap [13, 19–21].

Another promising topology of electrical traction machine is the transverse flux machine (TFM) which has a 3D magnetic flux path perpendicular to the rotation of the rotor. Due to its special structure, the electrical loading and the magnetic loading are decoupled with each other. Therefore, large number of pole pairs and high torque density can be achieved with TFM [22–25].

It should be noted that the magnetic flux flows three-dimensionally in both topologies. As a result, the isotropic SMC are usually used to manufacture the soft magnetic components. In other words, both AFM and TFM are promising candidates for the application of SMC, which should be researched in this thesis.

1.3 Target and design specifications of the work

The mechanical and electromagnetic properties of SMC have undergone a significant development in the past few years. Therefore, the target of this thesis is to investigate and to evaluate the application of SMC in the novel designs of electrical traction machine. For this purpose, a suitable design of electrical traction machine should be developed. Then, it is compared with the current design of RFM based on the laminated electrical steel. In addition, a prototype should be built and measured to validate the results of finite element analysis (FEA).

The research is carried out based on the cooperation with the industrial partners – Daimler AG and PMG Füssen GmbH within the project "Conception and Analysis of an E-Machine based on the Soft Magnetic Composites".

The design specifications are the basis of electromagnetic design. In this work, the specifications are determined considering the typical and potential applications for passenger cars from the industrial partner. Currently, an RFM is used as electrical traction machine in the car. Meanwhile, it is also used as the reference machine for the comparison with the novel electrical machines. In summary, the available space, the PM and the drive inverter are specified in Table 1.2. In this work, a TFM with inner rotor should be designed for a fair comparison with the current RFM.

Parameter	Value	Unit
Active outer diameter	301	[mm]
Active inner diameter	209	[mm]
Active axial length	75	[mm]
Axial length of air gap	1	[mm]
Maximal phase current	300	[Arms]
Maximal line to line voltage	167	[Vrms]
Mass of PM	0.83	[kg]
Remanence of PM	1.28	[T]
Coercivity of PM	1001	[kA/m]
Number of phases	3	-

Table 1.2. Design specifications of the electrical traction machine

1.4 Scientific contributions

The main scientific contributions of this thesis are listed as follows:

- Measurement of the magnetic properties including the magnetization curve and iron losses of different SMC products provided by PMG Füssen GmbH.
- Development of an iron loss model to calculate the iron losses in SMC parts based on finite element method (FEM). The model is validated with a test system comprising a yoke and a tooth made from the same SMC.
- Development of a 3D FEM simulation platform with integration of the measured data of SMC, a winding editor to calculate the important properties of double layer concentrated winding (DLCW) and a method to calculate the iron losses of AFM efficiently.
- Electromagnetic design of an AFM based on accurate equivalent magnetic network and 3D FEM.
- Development of a multi-objective design methodology of TFM based on Taguchi method, the nondominated sorting genetic algorithm 2 (NSGA2), the multi-objective particle swarm optimization (MOPSO) and fuzzy logic (FL).
- Comparison among the TFM, the AFM and the current RFM.

- Mechanical construction of AFM which is validated with the help of structural analysis and vibration analysis.
- Thermal analysis of AFM based on the accurate but time-consuming coupled thermal and computational fluid dynamics (CFD) analysis as well as the equivalent thermal network.
- Manufacturing and measurement of an AFM prototype.

1.5 Outline of the work

The doctoral thesis consists of nine chapters which reflect the progress of the research over time, as illustrated in Fig. 1.5.

Chapter 1 starts with an introduction of SMC and electrical traction machine in passenger cars. Subsequently, the targets and the design specifications of the research are presented. Finally, the main scientific contributions and the outline of this thesis are summarized.

Chapter 2 describes the classification of electrical machine depending on the main flux in the air gap and the relative direction between flux path and rotor rotation. Subsequently, the main topologies of PM excited AFM and TFM are described. At last, the most important literatures about the applications of SMC in RFM, AFM and TFM are reviewed and summarized.

A full characterization of SMC is the basis of FEA. **Chapter 3** deals with the measurement of the electromagnetic properties of different SMC including the magnetization curve and iron losses. On this basis, an iron loss model has been developed which is then used in the post-process of FEM to calculate the iron losses in electrical machine. In order to validate the model, a test system consisting of a yoke and a tooth made from the same SMC has been built.

In **Chapter 4**, the development of a 3D FEM simulation platform is described. First of all, the function and structure of the simulation platform are introduced. Subsequently, the most important components integrated into the simulation platform are explained in details. A winding editor to calculate the most important properties of double layer concentrated winding has been developed. Furthermore, an efficient method to calculate the iron losses of AFM based on limited FEA for $1/6$ electrical period is presented.

Based on the measurement of SMC and the developed simulation platform, the electromagnetic designs of an AFM and a TFM have been carried out and described in **Chapter 5** and **Chapter 6** respectively. Based on the comparison among the designed novel topologies and current RFM described in **Chapter 7**, it is concluded that the novel AFM is the most suitable structure for the application of SMC.

In order to validate the FEM results, a prototype of the designed AFM should be built and measured. For this purpose, a mechanical construction is proposed in **Chapter 8** which is very challenging due to the large axial force between the stator and rotor. On this basis, the strength and vibration analyses are performed to validate the mechanical construction. In addition, the temperature plays an important role in AFM because both adhesive and PM are sensitive to it. For this reason, the temperature is calculated based on the coupled thermal and CFD analysis. Subsequently, a lumped parameter thermal model has been developed to calculate the temperature efficiently with acceptable accuracy.

The manufacturing and measurement of the AFM prototype are described in **Chapter 9**. The experimental data including the back-EMF profiles, the losses, the torque, the informative efficiency map and so on are presented and compared with simulation. The possible reasons rendering the deviation between the measurement and simulation are discussed in this chapter as well.

Chapter 10 summarizes the major work carried out during the project. At last, the most relevant conclusions drawn from the previous work about the SMC, the topology of electrical machine and the future work are presented.

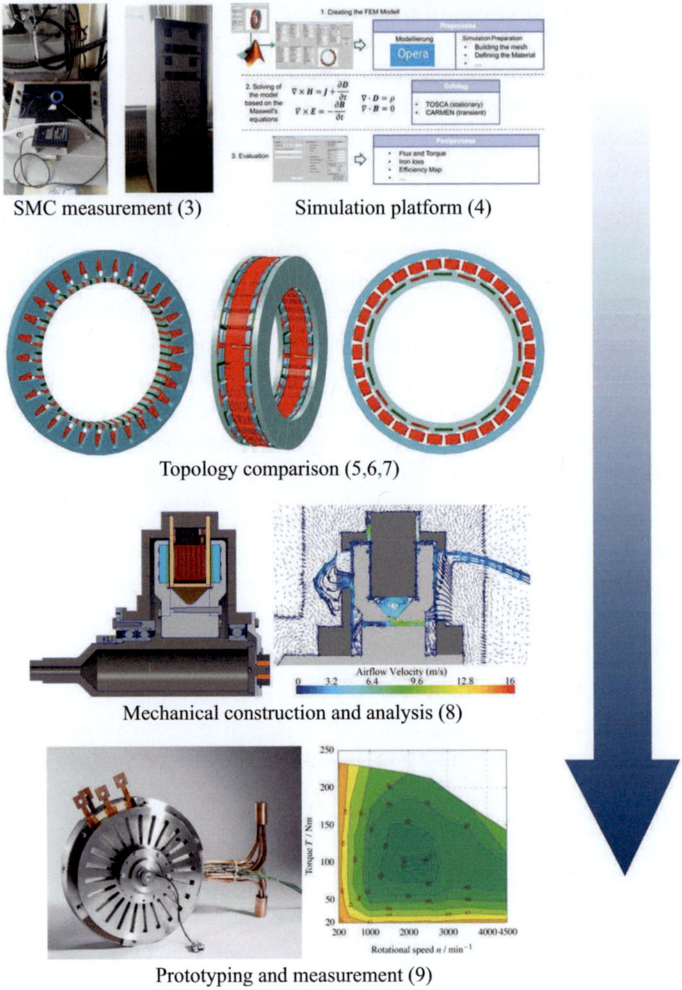

Figure 1.5. Structure of the doctoral thesis

2

Topology of Electrical Machine

In the past years, a lot of different topologies of the electrical machine have been developed. These topologies can be classified with many different methods. For instance, electrical machine can be divided into the direct current machine and the alternating current machine depending on the current waveform. Considering the existence of PM, electrical machine can be categorized as PM and electrically excited machine.

As illustrated in Fig. 2.1, the AFM is featured by the main magnetic flux flowing axially in the air gap, while the magnetic flux in RFM flows radially in the air gap. Furthermore, the main magnetic flux path is longitudinal to the rotational direction of the rotor in longitudinal flux machine (LFM), while the TFM has a magnetic flux path transverse to the rotor movement. In this thesis, two different kinds of electrical machines are investigated which are the longitudinal axial flux machine and the radial transverse flux machine respectively. For the sake of simplicity, the abbreviations AFM and RFM are used.

In this chapter, the classification of RFM and the reference machine for the comparison with novel topologies are introduced at first. Subsequently, the classical topologies of AFM and TFM are presented respectively. Thirdly, the mathematical model of PM excited synchronous machine based on the *d*-, *q*-coordinate system is presented which can be utilized to analyze the PM excited RFM, AFM and TFM. Finally, the most relevant literatures about the applications of SMC in the three different kinds of electrical machines are summarized.

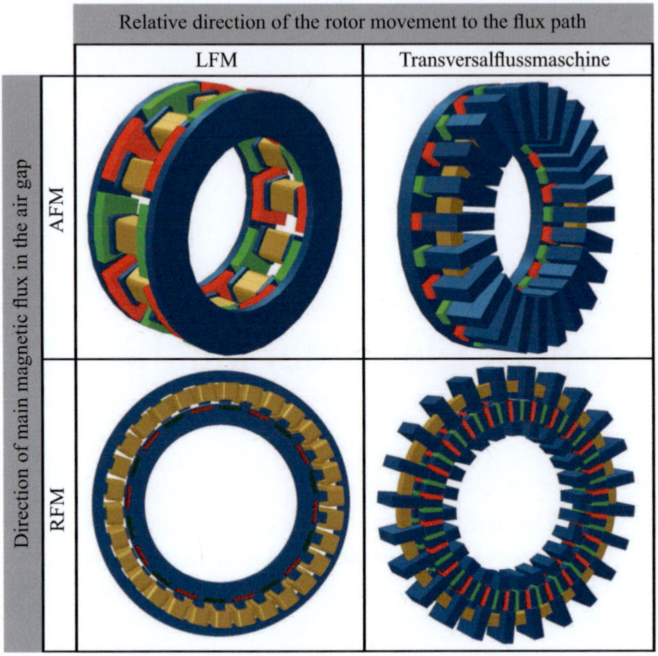

Figure 2.1. Classification of the electrical machines

2.1 Radial flux machine

The RFM can be classified depending on the current waveform. Compared with DC machine, almost only three-phase AC machine is used in the traction application. It can be categorized into asynchronous machine (ASM) and synchronous machine (SM) depending on whether the rotor speed equals the rotational speed of stator magnetic field. Nowadays, both PM excited RFM (PMRFM) and electrically excited SM have been used as traction machine in automotive industry.

In general, the PMRFM has higher torque and power density due to the application of the rare-earth PM with high energy density. In comparison, the ASM is more robust and cost-effective because of the absence of PM, which is especially suitable for the operation at high speed in field weakening area.

- RFM
  - DC-RFM
    - PM excitation
    - Electrical excitation
  - AC-RFM
    - Asynchronous
      - Squirrel-cage rotor
      - Slip-ring rotor
    - Synchronous
      - PM excitation
      - Electrical excitation

The RFM designed by Daimler AG is used as the reference machine for the comparison with other novel topologies, whose structure is illustrated in Fig. 2.2. In order to reduce torque ripple and vibration, the rotor consists of 7 segments which are rotated among each other for a small angle. The PMs are embedded in rotor, so that the eddy current losses are effectively limited compared with RFM, where PMs are mounted on the surface of rotor yoke. On the other hand, the reluctance torque is generated in this structure, which is of great importance when the RFM works at high speed with large field weakening current i_d.

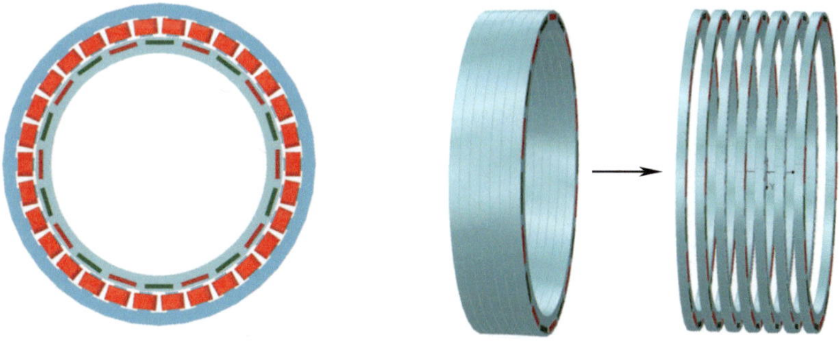

(a) FEM model of RFM

(b) Exploded view of Rotor of RFM

Figure 2.2. The current electromagnetic design of RFM

2.2 Axial flux machine

Although the widely accepted RFM becomes nowadays the dominant configuration, the AFM proves to be a quite competitive alternative, especially when the axial length of the free space is limited and simultaneously high torque and power density are required.

Compared with the axial flux induction machine [19, 26, 27] and the axial flux reluctance machine [28–30], the PM excited AFM (PMAFM) is more attractive because of its high torque and powder density, adjustable air gap, low vibration and noise, high efficiency and compactness. The greatest difficulties with PMAFM are the lack of fabrication technology of stator core and the large axial magnetic attraction force between stator and rotor. Thus, the construction and assembly of PMAFM is a great challenge.

Considering the different number of air gaps, various arrangements of PM, diverse winding configurations, existence of armature core and slots, the PMAFM can be further classified into many configurations with different performance.

- PMAFM
  - Single-sided
    - Slotted
    - Slotless
    - Coreless
  - Double-sided
    - PMAFM-IR
      - Slotted
      - Slotless
    - PMAFM-IS
      - PMAFM-IS-NN
        - Slotted
        - Slotless
      - PMAFM-IS-NS
        - Slotted
        - Coreless
      - PMAFM-ISA
  - Multi-stage

2.2.1 Single-sided PMAFM

The single-sided PMAFM (SS-PMAFM) has the simplest construction which comprises one stator and one rotor. For instance, two SS-PMAFMs are illustrated in Fig. 2.3 which comprise the double-layer concentrated winding (DLCW) and the distributed winding respectively. Rotors of both SS-PMAFMs are the same, where the adjacent PMs with red and green colour have opposite magnetization direction.

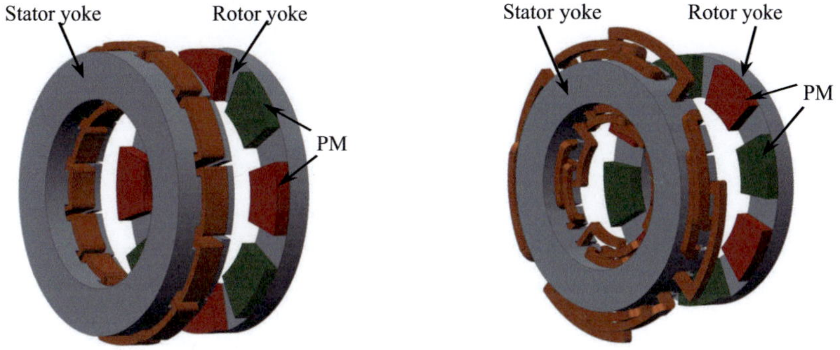

(a) Concentrated winding (b) Distributed winding

Figure 2.3. PMAFM with concentrated and distributed winding

Both distributed winding and concentrated winding can be applied in the PMAFM. Despite more space harmonics contained in the air gap field, the concentrated winding is more often utilized in PMAFM because of its short end winding, low copper losses and simplified manufacturing. Furthermore, there is no overlapping among the coils, so that the phases are physically separated leading to high fault toleration.

In comparison, the distributed winding can offer much fewer harmonics, which makes it the only considerable candidate for the axial flux induction machine. It should be noted that high magnetic attractive force exists in SS-PMAFM, which imposes high requirements on the bearing system.

2.2.2 PMAFM with an internal rotor

Compared with SS-PMAFM, the magnetic force on the rotor is obviously limited in the double-sided PMAFM with an internal rotor (PMAFM-IR). Two PMAFMs with slotted and slotless stator are illustrated in Fig. 2.4. For both machines, the main flux flows axially through the rotor. In addition, the non-magnetic and non-electric conductive materials are necessary to fix the PMs.

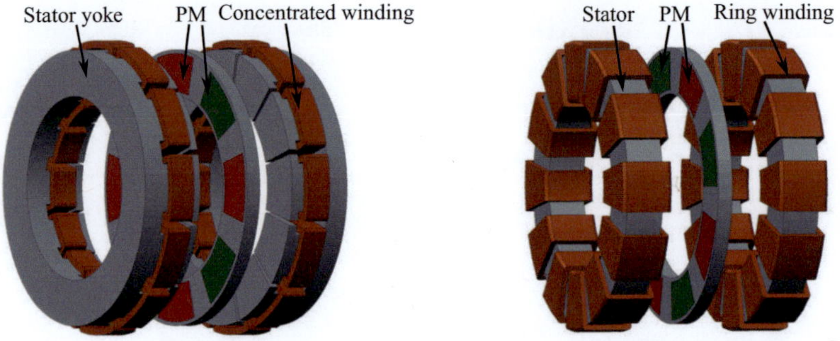

(a) Slotted stator

(b) Slottless stator

Figure 2.4. PMAFM-IR with slotted and slotless stators

Generally, the PMAFM with slotted stator has smaller air gap length and thus higher flux density. However, the stator teeth can cause local magnetic saturation, higher torque ripple, more space harmonics and higher eddy current losses in PMs. In comparison, the PMAFM with slotless and toroidal stator can eliminate the cogging torque, flux ripple and manufacturing costs resulted by the slots. Nevertheless, the disadvantages should also be taken into account. Firstly, the air gap becomes larger so that the flux density and torque become lower. In addition, the generated heat in the stator core cannot be effectively dissipated. At last, plenty of eddy currents can be induced in the windings because they are directly exposed to the PMs.

Different from the above two PMAFM-IRs where the magnetic flux flows axially through the PMs in the rotor, the main magnetic flux can flow circumferentially along the rotor yoke as well. In this case, a thick rotor yoke is necessary to avoid high magnetic saturation and reduction of torque density.

As illustrated in Fig. 2.4 and Fig. 2.5, the PMs can be either mounted on the surface of rotor yoke or buried in rotor, which can result in totally different characteristics.

(a) Surface mounted PM (b) Buried PM

Figure 2.5. PMAFM-IR with surface mounted and buried PMs

In general, the PMAFM with surface mounted PM has large effective air gap length and can offer higher peak torque density. However, this kind of PMAFM isn't suitable for high speed operation. On one hand, high current is needed to weaken the PMs at high speed. Simultaneously, a large amount of eddy currents can be induced in PMs. On the other hand, the mechanical stability becomes critical due to the increasing centrifugal force, because the PMs are normally glued on the rotor yoke.

In comparison, the PMAFM with interior PMs is more suitable for high speed operation due to its outstanding mechanical integrity at the cost of more complex construction and manufacturing. In addition, the eddy current losses and the demagnetization risk of PM can be effectively reduced as well.

2.2.3 PMAFM with an internal stator

Compared with PMAFM-IR, the attractive force on stator can be counterbalanced by placing it between two rotors. In many literatures, the double-sided PMAFM including an internal stator (PMAFM-IS) sandwiched between two external rotors with surface mounted PMs is named as the TORUS machine or yokeless and segmented armature machine (YASA).

Furthermore, two structures of PMAFM-IS can be developed based on the different arrangements of PM polarity and flux direction. In PMAFM-IS-NS, the aligned PMs on the two rotors have the same magnetization direction, while the aligned PMs on the rotors of PMAFM-IS-NN have opposite magnetization directions.

For instance, two PMAFM-IS-NNs are illustrated in Fig. 2.6. For the PMAFM-IS-NN with slotless stator, the main magnetic flux enters the stator and flows circumferentially in the stator disc. For the PMAFM-IS-NN with slotted stator, the main magnetic flux flows firstly axially in the stator teeth and then circumferentially in stator yoke. Considering the total magnetic flux from the two rotors, the stator yoke of both machines should be thick enough to avoid high saturation, which leads to higher weight and longer end windings.

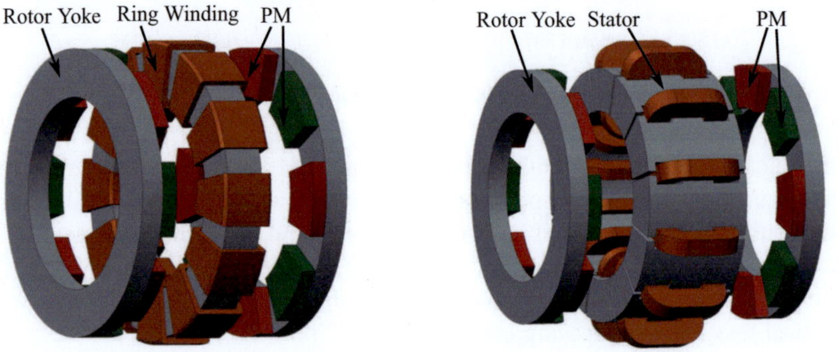

(a) Slotless stator (b) Slotted stator

Figure 2.6. PMAFM-IS-NN with slotless and slotted stator

In addition, the end windings in slotless PMAFM-IS-NN are much shorter resulting in less copper losses. The other differences between the slotted and slotless stators such as flux ripple and tooth saturation are already discussed in the PMAFM-IR which are also valid for the PMAFM-IS. In conclusion, the slotted PMAFM-IS-NN has higher torque density, more robust and yet more complex structure than the slotless PMAFM-IS-NN.

In comparison, two PMAFM-IS-NSs are illustrated in Fig. 2.7 where the main magnetic flux travels axially through the stator. Because the stator yoke of the slotted PMAFM-IS-NS is only used to connect the stator teeth together mechanically, the stator thickness is much smaller than the slotted PMAFM-IS-NN in

Fig. 2.6. The stator of the coreless PMAFM consists of only poly-phase windings which are normally cast together with resin to improve mechanical integrity and heat dissipation ability. Because of the absence of stator core, the coreless PMAFM-IS-NS machine has a short axial length. Besides, the iron losses and magnetic saturation in the stator are also eliminated so that a high efficiency can be achieved.

Similar to the slotless PMAFM, the cogging torque and torque ripple of coreless PMAFM are very low. The disadvantages with the coreless PMAFM are the high eddy current losses generated in the conductors and large effective air gap length which cause low armature inductance and low torque density.

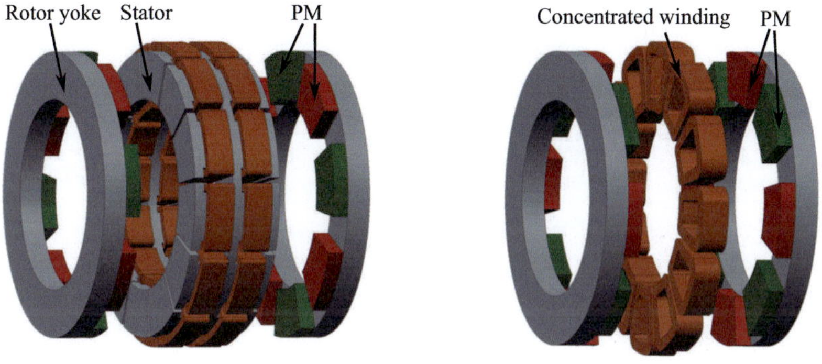

(a) Slotless stator (b) Coreless stator

Figure 2.7. PMAFM-IS-NS with slotted and coreless stator

Different from the above PMAFM-IS with surface mounted PMs, the PMAFM-IS can also be designed with interior or buried PMs to improve the mechanical robustness by protecting PMs against centrifugal forces. Moreover, higher torque can be delivered in the field weakening area due to the reluctance effect. The accompanied drawback is the manufacturing difficulty. Compared with the PMAFM-IR, the PMAFM-IS has less stator core material and copper causing lower iron and copper losses and higher power density. Because the stator is placed between two rotors, the contact surface between the stator and the ambient is much smaller than that of the PMAFM-IR. For this reason, the heat generated in the stator cannot be effectively dissipated.

As an important candidate for the application of SMC, the PMAFM with internal segmented armature (PMAFM-ISA) illustrated in Fig. 2.8 is a further development of the traditional PMAFM-IS topology which is often known as yokeless and segmented axial flux machine (YASA) or segmented armature torus machine (SAT). In this topology, the stator yoke isn't needed to conduct the magnetic flux. The PMAFM-ISA consists of two external rotors with surface mounted PMs and an internal stator comprising several independent segments with concentrated winding. Due to the absence of the stator yoke, the stator weight and the iron losses are obviously reduced, and the efficiency is improved.

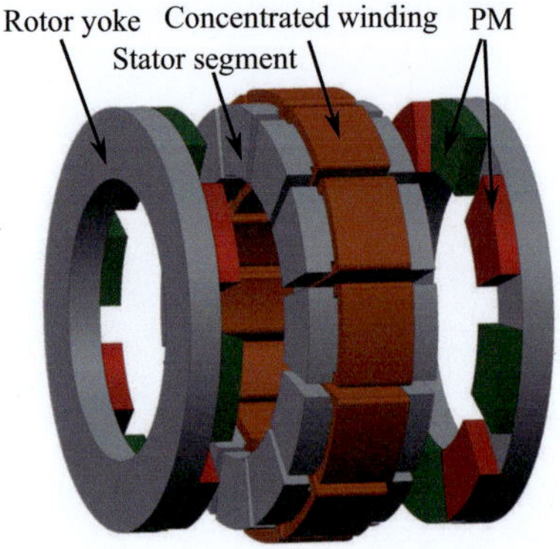

Figure 2.8. PMAFM-ISA

2.2.4 Multi-stage PMAFM

Because of the planar air gap, the multi-stage PMAFM (MS-PMAFM) can be easily realized by axially merging the single-sided PMAFM, the PMAFM-IR or PMAFM-IS together. For instance, the Fig. 2.9 shows a MS-PMAFM that consists of four stators and three rotors with interior PM.

The MS-PMAFM can be very attractive for certain applications where large power density is required, and simultaneously the external diameter is limited. Obviously, the construction is more difficult due to its high mechanical stress and complex structure. In addition, the costs of the machine might be high because of the utilization of large amount of PMs.

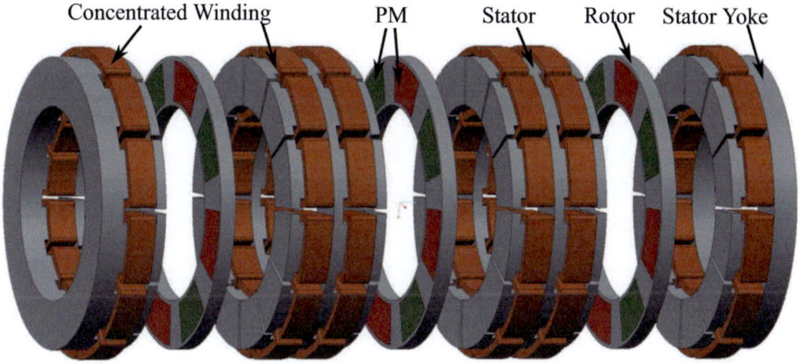

Figure 2.9. Multi-stage PMAFM

2.3 Transverse flux machine

As mentioned above, the electrical machine can be categorized into LFM and TFM considering the relative direction between the rotor movement and the magnetic flux path. In LFM, the main magnetic flux path lies within the same plane as the rotor movement while the magnetic flux in the TFM flows in planes that are perpendicular, in other words transverse to the direction of the rotor movement. For a better view, part of the basic PM excited TFM (PMTFM) is linearized and illustrated in Fig. 2.10.

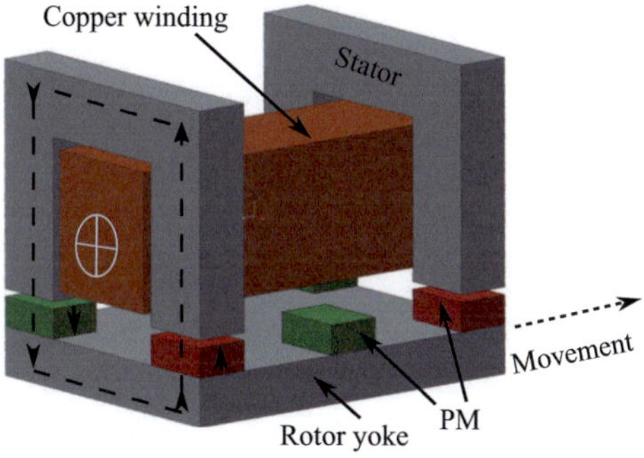

Figure 2.10. Part of a linearized one-phase PMTFM

The PMs with alternating magnetization direction indicated with the green and red color respectively are mounted on the surface of the rotor yoke. The stator consists of many separate cores which are used to conduct the magnetic flux from PMs indicated by the dotted line. The winding with simple structure and high filling factor is placed in the slots of the stator and fed with alternating current. The magnetic flux flows through the PM stator cores, the air gaps, the PMs and the rotor yoke.

Because of the special structure, the TFM has some outstanding advantages. Some of them are summarized as follows:

- High torque density
- Simple winding structure
- Decoupling of the magnetic loading and the electrical loading

Due to the special structure of TFM, the magnetic and electrical circuits are decoupled, which do not compete for the same space as in the LFM. Thus, a large number of poles can be achieved without increasing the external diameter, and the electromagnetic design is simplified. In addition, the winding used in TFM has usually simple structure and high filling factor which lead to lower copper losses and higher rated electrical loading. In conclusion, TFM can achieve high efficiency and outstanding torque density through increasing the number of poles and electrical loading.

Despite the above advantages, there are two considerable drawbacks preventing the wide applications of TFM:

Low power factor

The power factor of TFM is generally low because of the large amount of leakage flux which comprises the slot leakages, the leakages flux between the phases, the end winding leakages and the fringing fluxes. It is caused by a large number of isolated components in the magnetic circuit and many sides directly facing the air. In addition, the power factor reduces with increasing pole number. The reduced power factor causes a higher rating and costs of the drive inverter. As a result, for a given dimension, a high number of poles leads to small pole pitch, high leakage flux and consequently low power factor. Therefore, a compromise has to be made between the torque and the power factor.

Complex mechanical construction

The mechanical structure of the TFM is more complicated than the conventional RFM because a large number of isolated and small components are needed to form the magnetic circuit. As a result, the manufacturing costs and assembly difficulties are enhanced, and the mechanical integrity becomes worse.

It should be noted that at least two phases are needed to guarantee the continuous operation of TFM. For instance, a three-phase TFM can be obtained by stacking three one phase TFMs together along the axial or peripheral direction. Furthermore, either the PMs or the stator cores must be shifted to achieve a 120 electrical degree among the phases. For the TFM in Fig. 2.11, the stator cores are shifted while the PMs are aligned.

In order to reduce the leakage flux among the phases, a small gap or barrier between the cores of adjacent phases are necessary. Because of the electromagnetic independence among the phases without considering the leakage flux, it is only necessary to analyze one pole pair model of one phase.

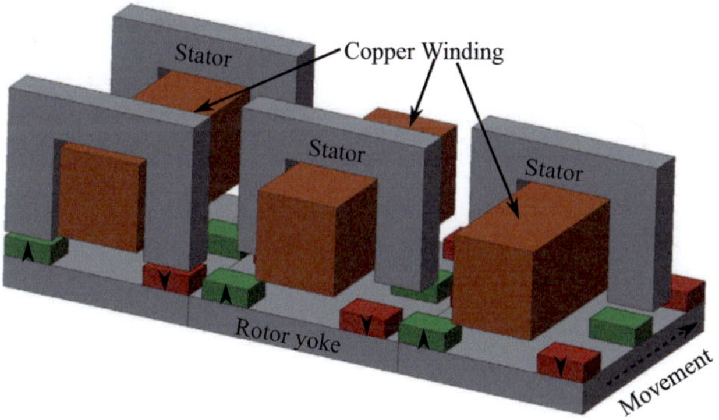

Figure 2.11. Part of a linearised three-phase PMTFM

In general, the TFM can be classified considering the type of excitation (electrical or PM excitation), the arrangement of the PMs (surface mounted or flux concentration), the placement of PMs (positive or passive rotor), the position of rotor (inner or outer rotor), the distribution of stator segments (single or double sided stator) and the configuration of the magnetic circuit (axially or circumferentially shifted phases).

- TFM
  - Reluctance TFM
  - PM excited TFM
    - Flux concentration
      - Single-Sided
      - Double-Sided
    - Surface mounted PM
      - Single-Sided
      - Double-Sided

A reluctance TFM (RTFM) and a single-sided PMTFM are illustrated in Fig. 2.12 where all components of the stator are placed on one side of the rotor. One of the advantages of RTFM is the absence of PM, which results in low costs and better mechanical robustness. Generally, the RTFM has a rotor with large weight and can offer lower torque density compared with the PMTFM. Compared with the TFM in Fig. 2.10, the PMTFM in Fig. 2.12 can utilize the generated magnetic flux from all PMs at the cost of increased weight and reduced available winding space.

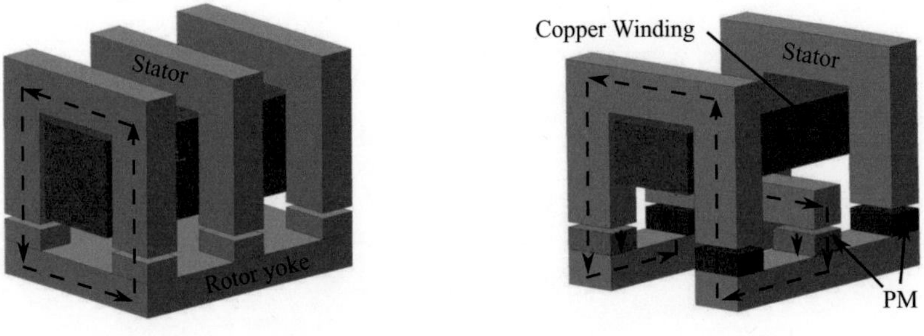

(a) RTFM (b) PMTFM

Figure 2.12. Reluctance TFM and TFM with surface mounted PM

A higher magnetic flux density in the air gap can be achieved by reducing the effective air gap and concentrating the flux from the PMs. For instance, a single-sided and a double-sided TFM with flux concentration are illustrated in the Fig. 2.13 where the PMs with inverse circumferential magnetization directions are embedded in the rotor.

Although a higher torque density can be realized with the flux concentration structure, the machines are difficult to be manufactured, and the mechanical integrity becomes worse, especially for the TFM with stator cores spread over both sides of the rotor. An obvious advantage of the TFM with twisted stator cores in Fig. 2.13 is that the stator cores are combined together forming a whole component for each phase, which reduces the constructional and manufacturing difficulties. A main drawback of the TFM with flux concentrating is higher leakage flux compared with the TFM with surface mounted PMs.

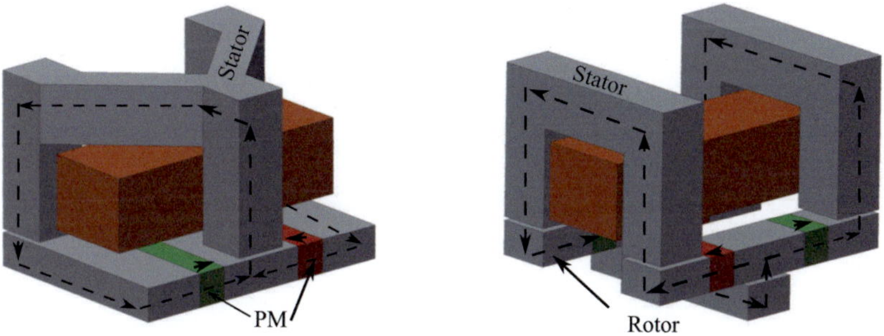

(a) Single-sided (b) Double-sided

Figure 2.13. Single- and double-sided PMTFMs with flux concentration

2.4 Modelling of PM excited synchronous machine

All the three topologies of electrical machine researched in this work are excited with PM which can be analyzed with the d-, q-coordinate system. The d-axis corresponds to the magnetization direction and q-axis is the axis possessing a deviation of 90 electrical degrees to the d-axis, as illustrated in Fig. 2.14.

In this case, the stator voltage u_d, u_q can be calculated from the phase resistance R_S, the current components i_d, i_q, the electrical rotational speed ω_el and the flux linkage Ψ_d and Ψ_q as follows [16]:

$$u_\mathrm{d} = R_\mathrm{S} \cdot i_\mathrm{d} + \frac{\mathrm{d}}{\mathrm{d}t} \cdot \Psi_\mathrm{d} - \omega_\mathrm{el} \cdot \Psi_\mathrm{q} \tag{2.1}$$

$$u_\mathrm{q} = R_\mathrm{S} \cdot i_\mathrm{q} + \frac{\mathrm{d}}{\mathrm{d}t} \cdot \Psi_\mathrm{q} + \omega_\mathrm{el} \cdot \Psi_\mathrm{d} \tag{2.2}$$

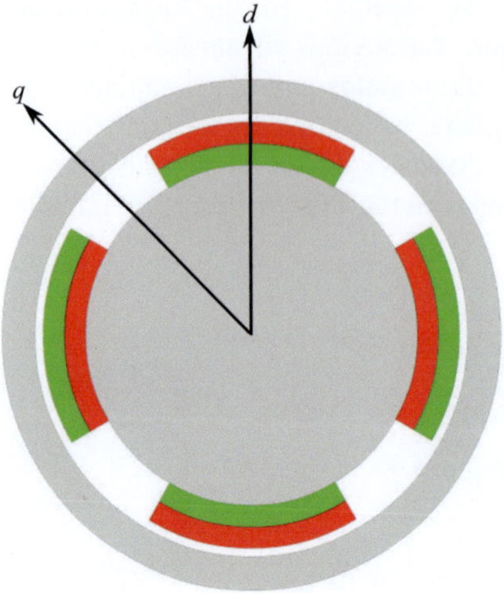

Figure 2.14. d-, q-coordinate system of a PMRFM

The ratio between the electrical rotational speed ω_{el} and the mechanical rotational speed Ω equals the number of pole pairs p. The rotational speed Ω is defined as the derivation of the rotor angle γ.

$$\omega_{\mathrm{el}} = p \cdot \Omega = p \cdot \dot{\gamma} \tag{2.3}$$

Furthermore, the flux linkage Ψ_{d} and Ψ_{q} can be determined from the stator currents i_{d} and i_{q}, the inductances L_{d} and L_{q} and the flux linkage generated by PM Ψ_{PM}.

$$\Psi_{\mathrm{d}} = L_{\mathrm{d}} \cdot i_{\mathrm{d}} + \Psi_{\mathrm{PM}} \tag{2.4}$$

$$\Psi_{\mathrm{q}} = L_{\mathrm{q}} \cdot i_{\mathrm{q}} \tag{2.5}$$

On this basis, the internal torque can be calculated using (2.6) which consists of the synchronous torque and the reluctance torque [16].

$$M_{\mathrm{i}} = \frac{3}{2}\, p \left(\Psi_{\mathrm{d}}\, i_{\mathrm{q}} - \Psi_{\mathrm{q}}\, i_{\mathrm{d}}\right) = \underbrace{\frac{3}{2}\, p\, \Psi_{\mathrm{PM}}\, i_{\mathrm{q}}}_{\text{sync. torque}} + \underbrace{\frac{3}{2}\, p \left(L_{\mathrm{d}} - L_{\mathrm{q}}\right) i_{\mathrm{d}}\, i_{\mathrm{q}}}_{\text{reluctance torque}} \tag{2.6}$$

2.5 State of the art

In order to take the advantages of SMC and to overcome the disadvantages, a lot of researches have been done in the past few years. Encouraging improvements have been achieved which enable a better understanding and improvement of SMC, an improved production technique and productivity [31, 32], a more accurate measurement and modelling of iron losses [33–36].

Because of the performance improvement, the SMC attract great interests in the field of electrical machine. It has been shown, new competitive electromagnetic designs of electrical machines can be realized by utilizing the unique features of SMC. Subsequently, some import applications of SMC in RFM, AFM and TFM are summarized and presented.

2.5.1 Application of SMC in RFM

The applications of SMC in almost all kinds of RFM have been researched. It has been shown that a direct replacement of laminated steel core with SMC core leads usually to poorer performance due to the disadvantages of SMC. More attentions should be paid to the electromagnetic design, the manufacturing and the assembly of electrical machine so that the advantages of SMC can be exploited to achieve an improved performance. Some important literatures depending on the topologies of RFM are summarized in Table 2.1.

Machine topology	Literatures
DC motor	[37–39]
Universal motor	[40–44]
Reluctance motor	[45–51]
Induction motor	[45, 52–60]
PM excited RFM	[12, 45, 61–76]

Table 2.1. Application of SMC in RFM

One of the possibilities to make use of the manufacturing flexibility of SMC is to design an armature with overhangs, as illustrated in Fig. 2.15. The special geometry leads to an increase of air gap axial length, a better utilization of PM and an improvement of torque density [38, 39, 44].

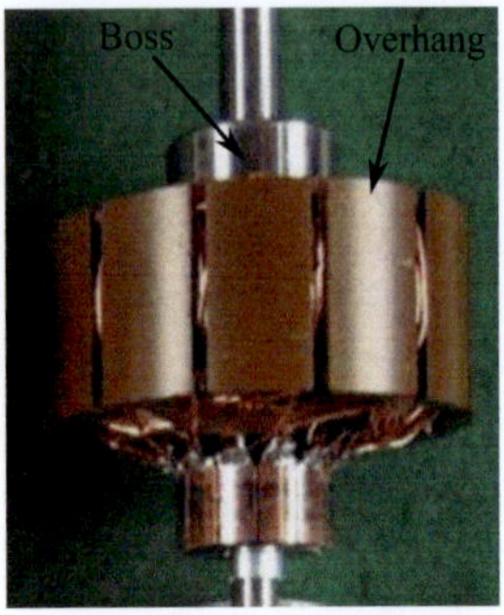

(a) Armature of a PM excited DC motor [38]

(b) Electrical excited DC motor [44]

Figure 2.15. Armature of a DC motor made from SMC with overhang

The design flexibility and isotropy of SMC are further exploited in [41]. As shown in Fig. 2.16, the stator core is made of SMC and separated into poles and two half-yokes which enables an application of bobbin winding and realizes significantly simplified assembly.

Figure 2.16. Stator of an universal motor made from SMC [41]

Theoretically, the reluctance machine isn't suitable for the application of SMC because the high permeability of soft magnetic components is required to generate large reluctance force. However, it could also be interesting in some special cases, especially when the simplifications of manufacturing and assembly are desired. For instance, the stator of a new DC field multi-tooth switched reluctance machine is illustrated in Fig. 2.17.

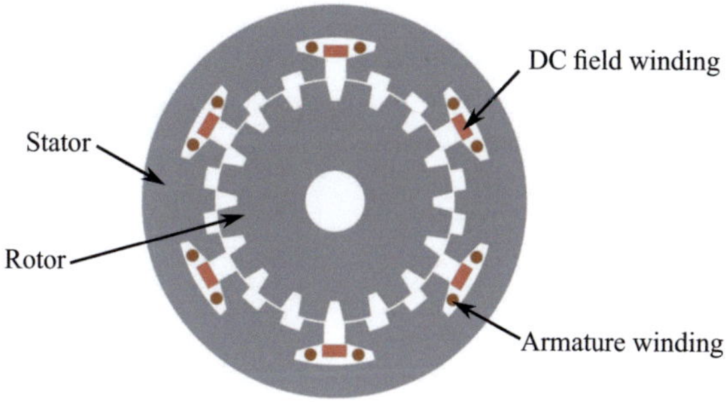

Figure 2.17. Stator of a switched reluctance machine made from SMC [41]

Similar to switched reluctance machine, soft magnetic material with high permeability is demanded in ASM. A direct replacement of the laminated electrical steel with SMC for induction machine leads to higher iron losses and lower efficiency, especially at high slip. However, some aspects of ASM such as manufacturing, thermal performance and recycling of copper can be improved through the application of SMC.

For instance, the SMC stator core of an ASM as traction motor is divided into many separate parts to enable an assembling of pre-formed coil [57, 59], as illustrated in Fig. 2.18. Furthermore, the tips of stator teeth are axially extended to enlarge air gap surface. The measurement shows that the efficiency under rated conditions and peak output power increase by 3.9 % and 5 % respectively. Furthermore, it should be noted that for the inverter-fed ASM with SMC stator core, the losses increased by the inverter harmonics are even lower than that of conventional AMS [53, 56].

In general, the PMRFM is more suitable for the application of SMC. It is less sensitive to the low permeability of SMC than the armature magnetized machines due to the large effective air gap caused by PMs. Furthermore, the isotropy and manufacturing merits of SMC can deliver additional considerable benefits. Thus, the utilization of SMC in PMRFM attracts a lot of attentions in the past time.

(a) Assembled stator (b) Yoke (c) Tooth(SMC)

Figure 2.18. Stator of an ASM made from SMC [57]

A PMRFM comprising an inner rotor with surface mounted PMs and an SMC stator with concentrated winding was reported in [67] where a number of design features can only be realized with SMC, as illustrated in Fig. 2.19. Firstly, the concentrated windings combined with the directly molded tooth with rounded axial ends can reduce copper losses, easier winding process and better thermal contacts. Besides, the armature core is segmented and split into core back sections and teeth, which allows the use of preformed coils with a high filling factor. In addition, the tooth tips and stator core are axially extended over end windings to reduce the stator back iron thickness and to enable a better use of PMs. The prototype has superior higher efficiency than the conventional machine, which indicates the feasibility and potential of SMC applied in low-cost and mass-produced PMRFM.

(a) Core components and coil (b) Assembled tooth (c) Assembled stator

Figure 2.19. Stator of a PMRFM made from SMC [67]

Another interesting design of PMRFM is presented in [75, 76] which is similar to a hybrid stepping motor, and the core components are made of SMC, as shown in Fig. 2.20. The axially magnetized PMs aren't mounted on the surface of the rotor but inserted between the SMC rotor cores. In addition, the complex SMC stator with multi-tooth is axially extended to reduce the copper losses. Compared with the PMRFM with a conventional laminated core, the machine shows higher efficiency at high speed.

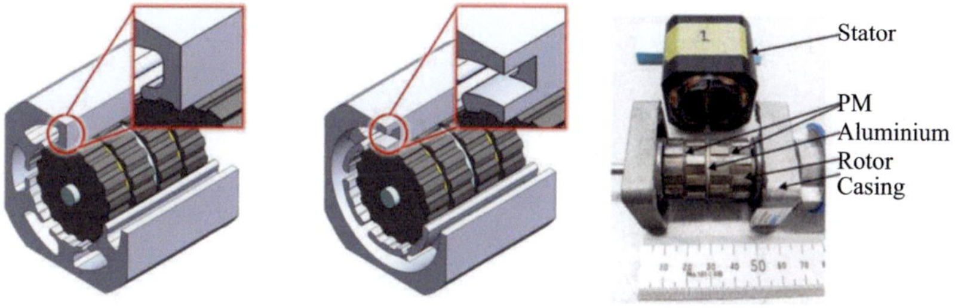

(a) Laminated core (b) SMC core (c) Prototype

Figure 2.20. A novel PMRFM with SMC cores [76]

All of the above machines are experimental prototypes developed in laboratory. However, some commercial applications of SMC in the electrical machines can also be found in market. An industrial servo motor was developed by ABB which achieves a size reduction by 1/3 and meanwhile improved performance compared with the previous design. Besides, a PMRFM with an inner SMC stator used in anti-lock braking system was developed by Aisin Seiki Co. Ltd. which realizes 36 % reduction in axial length and 17 % reduction in weight. The application of SMC in a DC pump motor was performed by Laing GmbH to achieve a lighter stator with simplified structure and less number of components.

2.5.2 Application of SMC in AFM

AFM has been first developed in 1830s [77–81]. Since then, it wasn't widely accepted as RFM because of the construction and assembly difficulties resulted mainly from the strong axial force between the stator and rotor.

	Slotted	Slotless	Coreless
SS-PMAFM	Water pump: [82], Wheel motor in scooter: [83], Wheel motor in wheelchair: [84], Wheel motor in solar powered vehicle: [85], Others: [86]*, [87]*, Servo motor: [88]*	Vibrator for mobile phone: [89]	Wind generator: [90, 91]
PMAFM-IR	Wheel motor in EV: [92–94], [95]*, Wind generator: [96]*	Direct-drive of scooter: [97, 98], Flywheel energy storage system: [99]	
PMAFM-IS	Wind generator: [100]*, [101]*, Wheel motor in HEV or EV: [21]*, [102]*, [103]*, [20, 104–107], Fan motor: [108], Industrial motor: [109], Others: [110]	Ship propulsion: [111], Pump motor: [105, 112], Automotive stater/alternator [113, 114], Direct-drive elevator: [115], Generator: [116, 117]	Generator: [118, 119], Flywheel energy storage system: [120], In-wheel motor: [121]
MS-PMAFM	Generator: [122]	Ship propulsion: [123], Direct-drive mixer: [124]	Wheel motor in EV: [125–127], Aircraft drive: [128]

Table 2.2. Various applications of PMAFM (*:SMC-based)

However, the PMAFM becomes an attractive candidate in the applications where high power and torque density, high efficiency and simultaneously short axial length are required. In the past time, a large number of researches have been carried out to investigate the application of different PMAFM for various applications which are summarized in Table 2.2.

Furthermore, the rated power and operating speed of some PMAFMs listed in Table 2.2 are illustrated in Fig. 2.21.

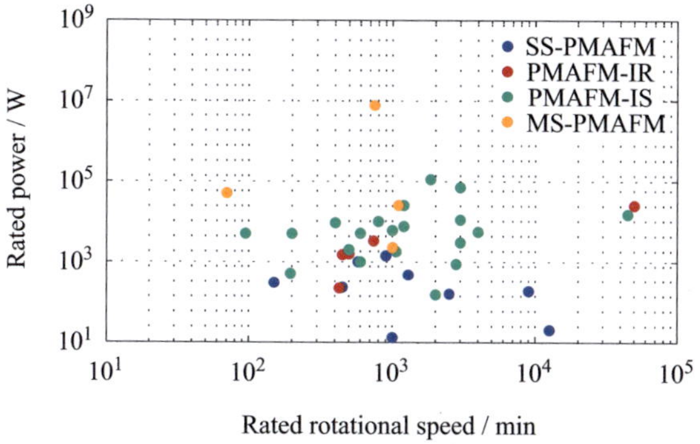

Figure 2.21. Rated power and speed of some PMAFMs

The most significant conclusions deduced from the performed researches in the past time are summarized as follows:

- Compared with PMRFM, the PMAFM with concentrated winding can deliver higher power density, better modularity, higher torque density and higher efficiency in the direct-drive applications with short axial length such as wind turbine and in-wheel motor of HEV or EV [105, 106, 109, 129–132].
- The electromagnetic design of PMAFM can be carried out based on the analytical method or FEM. Compared with analytical method, the analysis based on FEM is more accurate but requires much more computational time, especially in the case of 3D FEM [86–88, 107, 129, 133–139].
- The PMAFM with stator made from AMM or SI has higher torque density, lower losses and higher efficiency compared with the PMAFM with SMC stator possessing the same size which has however higher manufacturing difficulties and costs [79, 109]. The PMAFM with surface-mounted PMs is a suitable candidate for the application of SMC because of the 3D magnetic flux. Furthermore, it is less sensitive to the low permeability of SMC due to the large effective air gap length. Furthermore, the rotor yoke can be made from solid iron, electrical steel or SMC [88, 106, 110].

- The large axial force between the stator and the rotor of the PMAFM results in great difficulties in the construction and assembly [79].
- The surface-mounted PMs are often skewed and segmented to reduce the torque ripple and eddy current losses [140, 141]. Another method is the utilization of the magnetic wedges which are placed in the slots to form the closed or semi-closed slots [106, 110].
- The concentrated winding is usually utilized in PMAFM which can be realized with either one layer or double layer structure [122]. It is simpler to assemble and has shorter end-windings, higher filling factor and lower copper losses. In addition, the harmonics contained in MMF have dominant influence on the eddy current losses of the PMs [107, 142].
- In PMAFM-IS, the copper can be utilized for both surface of the stator. Besides, the end windings have reduced length compared with the PMAFM-IR, and thus the efficiency is improved [106]. Compared with PMAFM-IR, PMAFM-IS can achieve higher power density in high current density [143].
- The different topologies of PMAFM are compared in detail in terms of torque and power density, torque ripple, cogging torque, axial length, copper losses, filling factor and so on in [77, 140, 141, 144–148]. Among the classical topologies described in section 2.2, major efforts have been made and some novel topologies of PMAFM have be made to improve the capability of field control [149, 150], to improve the modularity and torque density [151], to develop novel design of winding with higher filling factor and higher allowable current density [148, 152], to achieve an iron free machine through Halbach array of PM [121, 148], and to reduce the cogging torque and torque ripple [153–158].

2.5.3 Application of SMC in TFM

The TFM was firstly proposed by Weh and May in 1986 [159–161]. Since then it has become an interesting research hot spot for many years until now. The TFM is suitable for the application of SMC because the main magnetic flux flows three-dimensionally. In addition, the magnetic components of TFM have normally complex form which are hard or even impossible to be manufactured with laminated steel. At last, the number of poles of TFM is always higher than that of conventional RFM, which corresponds to higher electrical frequency. Therefore, the application of SMC can effectively limit the eddy current losses.

During the past time, a lot of researches have been carried out to analysis different TFMs, and some prototypes have been manufactured for various applications. Some of them are summarized in Table 2.3.

	Single-Sided	**Double-Sided**
RTFM	Crankshaft starter-alternator: [162, 163], Direct Servo-drive: [164], Conveyor: [165, 166], Others: [167]*	Traction drive for a wheel loader: [168]*,[169]*, Exercise bike: [170], Others: [171]
Surface mounted PM PMTFM	Free Piston Energy converter: [172, 173], Wheel motor for EV: [174], Wind turbine generator: [175], [176]*, Generator: [177], Turret application: [178], Downhole application: [179], Direct drive application: [180]*, [181]*, High speed application: [182–184]*, Others: [185]*	Wind turbine: [186], Servo motor: [187], Direct-drive of EV: [188], Electrified propulsion of ship: [189]
Flux concentration PMTFM	Wind generator:[190, 191], Propulsion of EV and HEV: [192]*, Railway traction system: [193], In-wheel motor of hybrid bus: [24, 25], Wheel Hub Drive: [194]	Wave energy generator: [195], Ship propulsion: [196, 197], Electric and hybrid propulsion application: [198]*, Stocker system: [199], Electric-driven city bus: [200]*, [201], Generator: [202], Others: [203]

Table 2.3. Various applications of TFM (*:SMC-based)

Furthermore, the performance of some suggested TFMs are compared in terms of the torque density and power factor, as illustrated in Fig. 2.22.

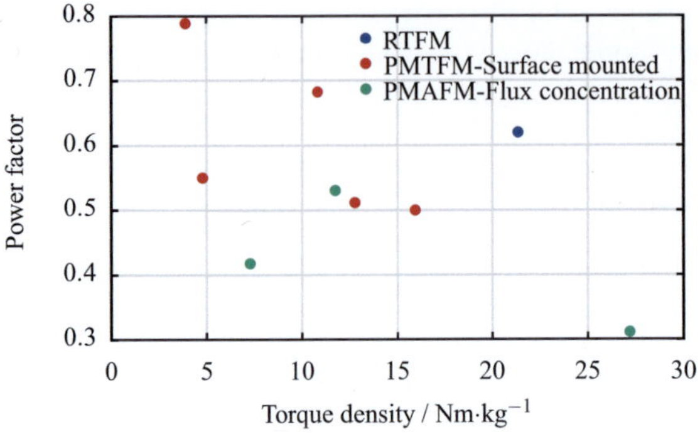

Figure 2.22. Torque density and power factor of some TFMs

Based on lots of detailed investigations and researches, some important conclusions about the TFM can be drawn which are summarized as follows:

- Compared with RFM, TFM is distinguished by its high pole number, high torque density, simple winding structure with high filling factor and low copper losses, high efficiency and improved fault tolerance due to decoupling of phases. For instance, the specific output torque of TFM is about 5$\sim$10 times of a standard industrial induction motor [204]. Therefore, the TFM is very appropriate for the high-torque and low-speed applications such as direct drive in vehicles or elevators [174, 178, 205]. In general, the ratio of the achievable power to total machine weight for TFM ranges between 0.5 $\sim$ 2 kW/kg [162]. Nevertheless, the disadvantages such as high leakage flux, complex structure and construction, high-cost manufacturing and assembly should also be taken into account.

- Compared with RTFM, the torque density of PMTFM is further improved by utilizing PMs, especially the rare-earth PMs with high energy density [162]. The performance of PMTFM is insensitive to the low permeability of SMC than RTFM as the magnetic reluctance of PM dominates the magnetic circuit. In general, the TFM with PM excitation can achieve higher force density by a factor of 3 to 5 compared with that of RTFM [199, 206]. However, the RTFM based on the reluctance effect is free of PM or electrical excitation, which leads to low costs, manufacturing simplicity, easy assembly and outstanding robustness [162, 207]. The magnetic field of RTFM can be conditioned by using magnetic shields such as PMs in order to enhance the torque density

[208–211]. The PMs can be placed either on the lateral sides of teeth, perpendicular on the lateral sides of the air gap or parallel with the air gap sides. The shielding of both stator and rotor is more efficient than shielding only one part, which however leads to more constructional and manufacturing difficulties.

- The components of stator can be placed on one side or both sides of the rotor. Compared with the single-sided PMTFM, the PM are better exploited in the double-sided PMTFM, and both air gaps contribute to the torque generation. However, the mechanical construction and assembly become more complicated because the additional cantilevered arrangement is necessary to support the rotor components [187].
- Different from the classical TFM whose stator comprises a large number of independent small parts, the stator of the PMTFM with claw-pole structure (PMTFM-CP) is a complete part and consists of half opened claw poles, within that the winding is housed. Because of the complex geometry, the magnetic flux flows truly 3D in the PMTFM-CP, which makes the application of SMC necessary. In conclusion, the PMTFM-CP can reduce the number of standalone components of stator, which leads to an improvement of the mechanical stability and a simplification of the manufacturing and assembly. But it should be noted that the overlapping of the stator teeth increases the leakage flux [187]. For instance, a PMTFM-CP with an inner rotor employing the surface mounted PMs was designed and prototyped in[212] to investigate its potential application as a high speed industrial servo drive. A comparison with the conventional PMRFM indicates that the newly developed PMTFM-CP has much higher peak and rated torque over the complete speed range.
- In addition, the PMs in PMTFM-CP can be either mounted on the surface of rotor yoke [182, 185] or buried in the rotor to form the flux concentrating structure [213]. The PMTFMs with different PM arrangements are compared in [214–216]. Compared with the PMTFM illustrated in Fig. 2.10, the PMs in the rotor of PMTFM with flux concentration can be completely utilized, which results in a reduced constructional complexity, a higher power factor and torque density, perhaps by the order of $20 \sim 50\%$ [204, 217]. However, it leads to a more complex structure, less robustness and an increase of magnet leakage flux of the rotor [176].
- The PMs can be either located on the fixed stator or the rotor of TFM. A single-sided and a double-sided PMTFM with passive rotor are illustrated in Fig. 2.23 which demonstrate better mechanical robustness because the PMs don't suffer from the centrifugal force [195]. In addition, the active water cooling of PMs

becomes much easier, which facilitates higher electrical loading. It should be noted that the improvements are achieved at cost of more PM material, mechanical complexity and manufacturing costs [218–220].

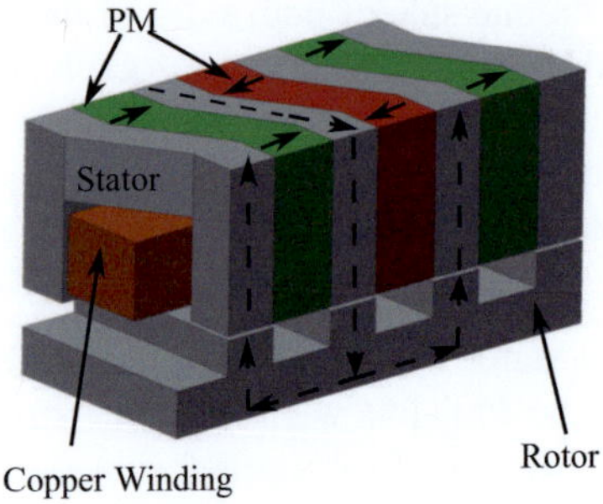

(a) Single-sided PMTFM

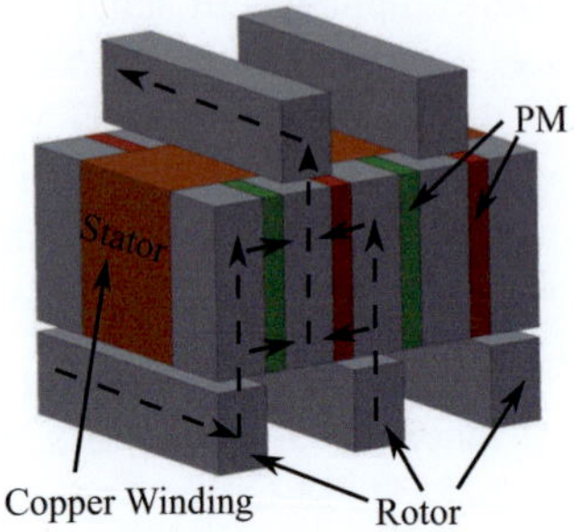

(b) Double-sided PMTFM

Figure 2.23. PMTFMs with passive rotor

- In principle, the phases of TFM are magnetically independent from each other. In fact, they are coupled due to the inter-phase leakage flux, which leads to reduction of average torque and higher cogging torque. For this reason, there should be enough free space among the phases. Sometimes, aluminium barriers or silicon steel sheets can be inserted between the phases as well to achieve a better decoupling [180, 192].
- The low power factor is a significant drawback of TFM which results in increasing copper losses and substantial over-rating of the drive inverter [176]. The power factor of TFM lies typically in the range of $0.35 \sim 0.55$ [173, 179, 214, 221]. The most important reason is the high fringing and leakage flux resulted by a lot of isolated components and large area facing air. The leakage flux can be reduced by increasing pole width and reducing the pole number, which in turn leads to reduction of torque density. For this reason, a compromise has to be made between power factor and torque density [222]. In [223], power factors of the PMTFM in Fig. 2.10 and the PMTFM with additional bridge in Fig. 2.12 are determined which are 0.52 and 0.47 for the original design and increase to 0.62 and 0.57 for the optimized design.
- Two different arrangements of phases of PMTFM are investigated in [224–226] which are the conventional axial phase arrangement and the in-plane arrangement respectively. As illustrated in Fig. 2.24, the phases are stacked

in axial direction or disposed next to each other in circumferential direction around the stator respectively. For the latter, the total axial length can be substantially reduced, and the torque density remains high [226]. Compared with the full winding construction, each phase can be divided into two halves and placed opposite to each other in the split winding structure to improve the mechanical balance and stability [225].

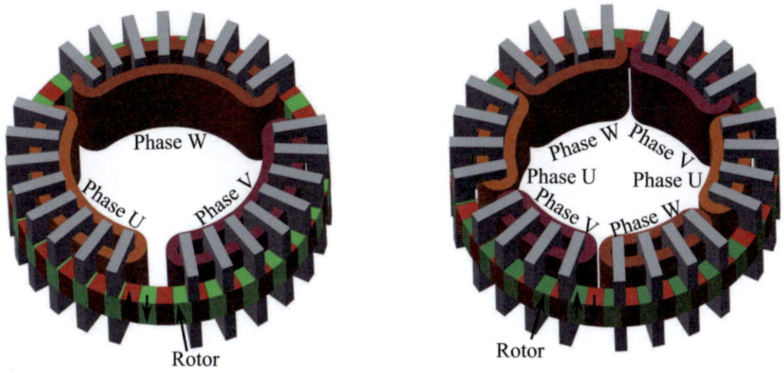

(a) Full winding (b) Split winding

Figure 2.24. PMTFM with in-plane arrangement

- For some TFM topologies, the soft magnetic parts can be made from either SMC or lamination. Because of its low permeability, the TFM with SMC parts needs higher ampere turns to achieve the same torque density compared with TFM based on electrical steel, which leads to higher copper losses [206, 226]. In addition, the mechanical properties of SMC are inferior to laminations, which leads to a more difficult construction [23, 25]. However, the SMC allow easier manufacturing with less costs, higher shaping flexibility and lower eddy current losses [176, 183]. Therefore, the SMC are the only possible candidate for the TFM with complex 3D flux components which are hard or even impossible to be manufactured with laminations [181]. Furthermore, the SMC and the laminated steel can be used in combination in PMTFM as well, as suggested in [167, 206, 224, 227]. As shown in Fig. 2.25, three PMTFMs with the same rotor but different stators made from SMC, laminations and a combination of both materials are compared in [228].

According to the experimental results, the PMTFM possessing stator made only from electrical steel with slits in teeth offers the highest output power while the PMTFMs with stator made from SMC and lamination have lower core losses.

(a) SMC (b) SMC and laminations (c) Laminations

Figure 2.25. PMTFMs with different stator [228]

Even for complex PMTFM-CP, combined usage of SMC and laminated steels are possible [229–231], as shown in Fig. 2.26. With this method, the performance of PMTFM can be improved due to reduced magnetic reluctance and iron losses. Through using core made from AMM instead of Fe-Si laminations, the iron losses can be further reduced, which leads to a higher efficiency.

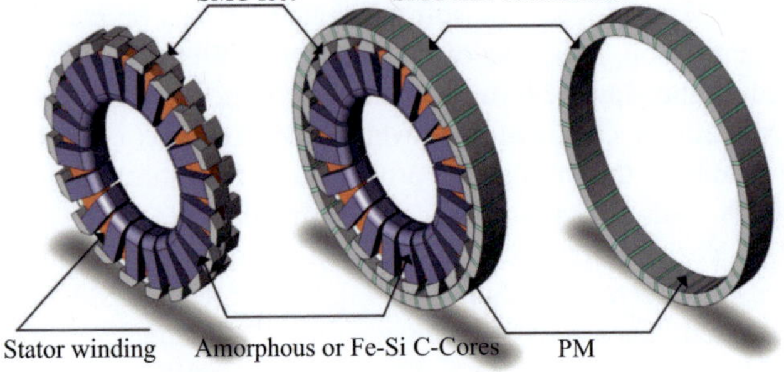

Figure 2.26. PMTFM-CP with stator made from SMC and electrical steel [231]

- Due to the complicated 3D flux components, design and optimization of TFM are very difficult and time-consuming. The performance analysis of TFM can be carried out based on the analytical method [221] or the 3D FEM [192, 207].

Most of the analytical calculations are based on the equivalent magnetic reluctance network and empirical formulas which have limited accuracy because of the complex 3D flux paths and large leakage flux [232]. Although the 3D FEM is extremely time-consuming, the most accurate prediction of machine performance can only be obtained through this method [233]. Furthermore, some efforts have also been made to investigate the TFM with 2D FEM instead of 3D FEM. It is found that the 2D FEM can be used for the qualitative investigation and parameter research [199, 219, 234].

- In the past time, a lot of novel topologies of TFM have been developed, and some of them are illustrated in Fig. 2.27. These topologies are developed aiming to increase the torque density by increasing the air gap area [203] or by using laminations instead of SMC [23], to improve the power factor [173, 186], to reduce the rotor leakage flux and to reduce the costs by using ferrite magnet instead of rare-earth PM [22], to reduce the cogging torque [192], to improve the fault tolerance [170], to improve the mechanical stiffness by using modern materials such as carbon-fiber reinforced plastics [168], to incorporate the merits of both the longitudinal and transverse flux machine [235], to replace the planetary gear in the HEV Prius [236], to increase the thermal dissipation ability by using hollow copper pipes with an internal coolant flow as windings in the stator [169] or to simplify the assembly and manufacturing by replacing many small PMs with a simple magnet ring placed between two rotor cores in the hybrid PMTFM [237–240].

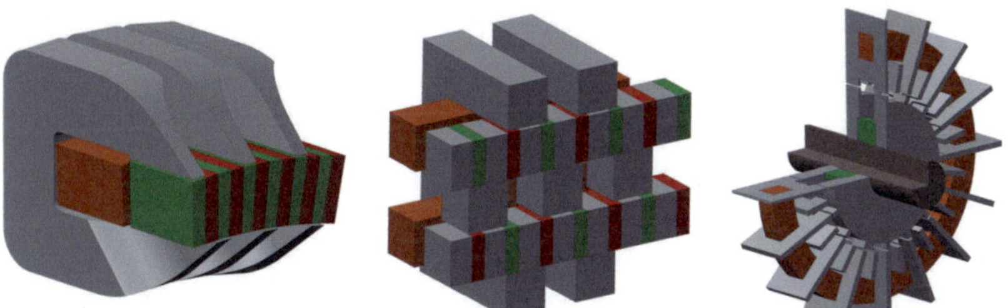

(a) C-shaped stator cores [196] (b) E-shaped stator cores [203] (c) Hybrid-PMTFM [237]

Figure 2.27. Novel topologies of TFM

As recent commercial application of SMC in TFM, the Twinbird Corporation in Japan has built a double sided PMTFM-CP for a linear compressor which results in a significant reduction of components number, manufacturing costs, weight and noise. In addition, the LYNG Motor has developed a two-phase PMTFM-CP for the valve drive application to improve the accuracy and reliability. Both TFMs are illustated in Fig. 2.28.

(a) TFM of Twinbird corporation (b) TFM of LYNG Motor

Figure 2.28. Commercial applications of two TFMs based on SMC

3

Measurement of SMC

In comparison with laminated electrical steel, manufacturing process of SMC is substantially simplified. However, there are still many factors which can influence the properties of SMC. These factors should be considered at the beginning of design process. As the basis of electromagnetic design, magnetic properties of SMC should be measured with specimens conforming to standards. Furthermore, iron loss plays an important role for electrical machine. Thus, an iron loss model based on 3D FEM has been developed and validated.

This chapter is organized as follows. First of all, the basic knowledge about the manufacturing of SMC and the main factors determining SMC properties are introduced. Subsequently, the measurements with the ring specimens made from different kinds of SMC are carried out in order to determine their magnetization curves and iron losses with different frequency and flux density. In addition, a model to calculate iron losses based on 3D FEM has been developed, whose coefficients are determined based on the measured iron losses through curve fitting. In order to validate the developed iron loss model, a simple test system consisting of a yoke and a tooth made from the same SMC Siron®STestb has been constructed and excited with sinusoidal voltage. The total losses of the complete system are measured and calculated with 3D FEM respectively.

3.1 Basic knowledge of SMC

The SMC are made of a great amount of fine metallic particles coated with a thin insulation layer. The manufacturing process consists of only three steps, namely the mixing, compaction and heat treatment, as illustrated in Fig. 3.1.

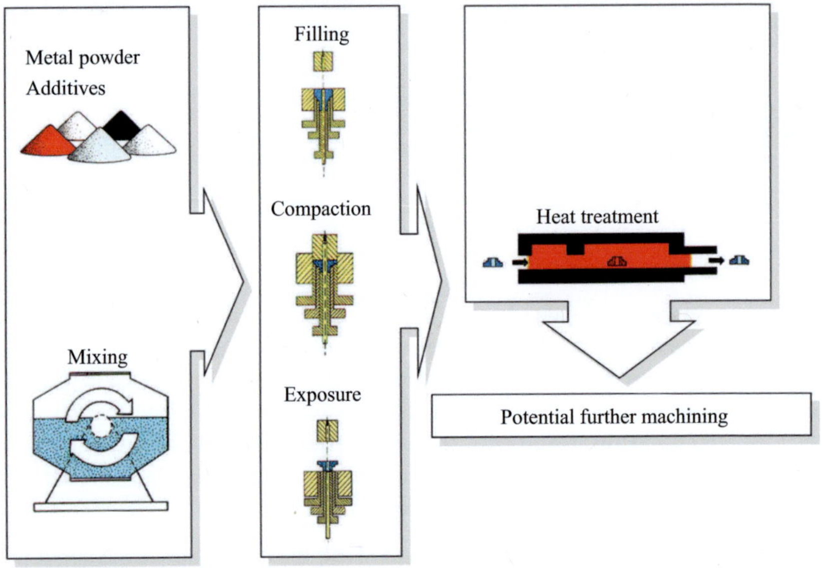

Figure 3.1. Manufacturing process of SMC

At first, extremely pure and highly compressible metal powder, insulation material, lubricant and sometimes binder material are homogeneously mixed and poured into a die which are then directly compacted into the desired shape and dimensions at high pressure. Subsequently, the ejected components are heat-treated to anneal and to cure the bond, so that the residual mechanical stress can be partly relieved and the magnetic performance can be improved.

Owing to the above manufacturing process, magnetic parts with complex geometry, high surface quality and good tolerance can be manufactured. In addition, waste of raw material is very limited during the manufacturing, which leads to a high level of material utilization.

Depending on the properties of the raw material and the manufacturing process, a wide range of magnetic and mechanical properties of SMC can be achieved. The most important factors that can determine the SMC performance are summarized in Fig. 3.2.

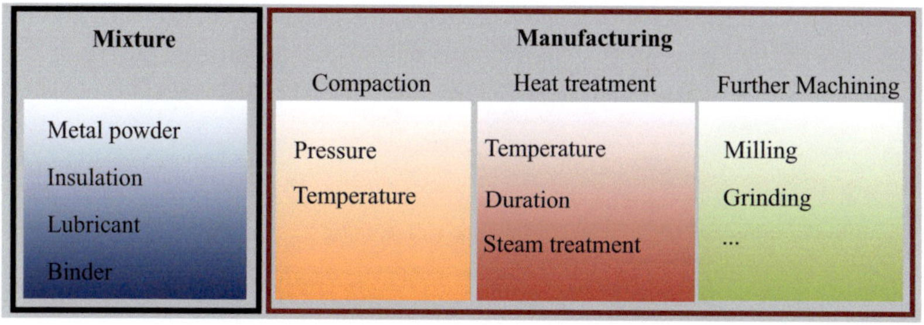

Figure 3.2. Dominant factors of SMC properties

For instance, the size of base particles can effectively influence the SMC properties. Two different kinds of SMC are illustrated in Fig. 3.3 which are developed for the application in electrical machines and power electronics respectively.

(a) SMC for electrical machine (b) SMC for power electronics

Figure 3.3. SMC with different particle size [KIT IAM-KM]

In summary, an increasing particle size results in lower electrical resistivity, higher eddy current losses and higher permeability because the movement of domains is less hindered. For this reason, the finer powder is suitable for high frequency application while the coarser particles yield better performance at low frequency domain. The influences of other factors in Fig. 3.2 are explained in [36] in details.

3.2 Measurement with SMC ring

The electrical and magnetic properties of different SMC including magnetization curve and iron losses under alternating field with different flux density and frequency are measured with ring specimens. As illustrated in Fig. 3.4, the SMC rings have the same geometric dimensions whose outer diameter D_a, inner diameter D_i and height h equal 55 mm, 45 mm and 5 mm, respectively.

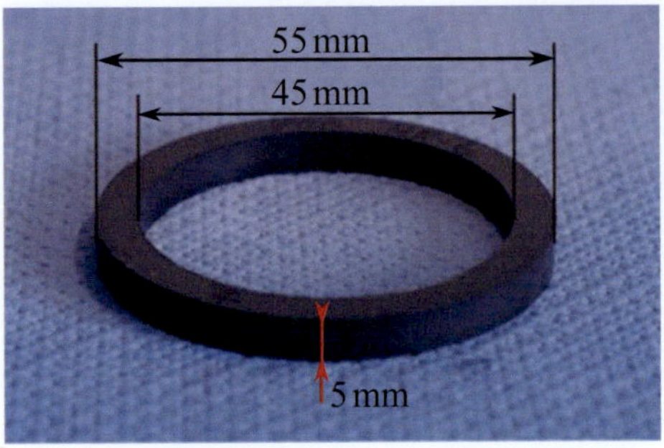

Figure 3.4. SMC ring specimen

The ring specimens are directly compacted and they are distinguished from the mixture, compaction and heat treatment. The measured results and some other important properties such as thermal conductivity and mechanical strength delivered by the industrial partner are listed in Table 3.1.

Specimen	**Density** [g/cm^3]	**Machin-ability**	**TRS** [MPa]	**E-Module** [GPa]	**Thermal conductivity** [W/mK]
Somaloy 700HR3P	7.51	×	91	177	15
Siron® S720	7.49	×	120	135	42
Siron® S280b	7.40	✓	90	110	35
Siron® S300b	7.30	✓	120	150	38
Siron® S400b	7.27	✓	135	130	42
Siron® STestb	7.40	✓	105	137	40

Table 3.1. Some important properties of SMC [KIT IAM-KM & PMG Füssen GmbH]

It should be noted that the first two kinds of SMC do not contain binder material which cannot be machined after the compaction. Furthermore, the steam treatment is performed for both materials to improve the mechanical property.

3.2.1 Electrical resistivity

The electrical resistivity of different SMC material can be measured with ring specimen based on the four-point method. As shown in Fig. 3.5, the circuit is supplied with a DC current source, and the total current i flowing in the SMC ring along two parallel paths is measured with an ammeter.

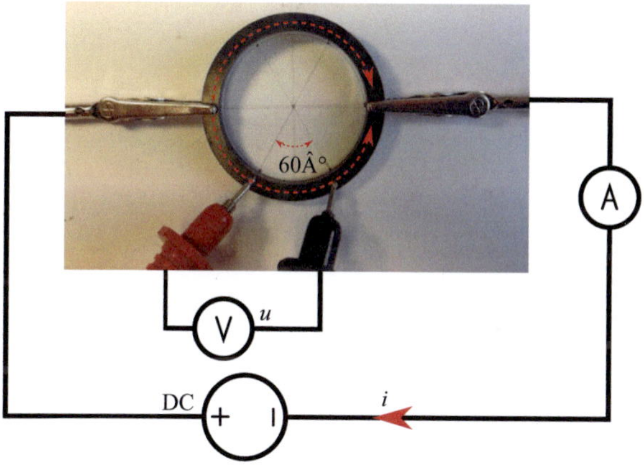

Figure 3.5. Resistivity measurement with four point method

With the measured voltage u and the geometric parameters $D_\mathrm{o}, D_\mathrm{i}, h$, the electrical resistivity can be determined using (3.1). The A and l_mag are cross sectional area of ring specimen and length of the curve along the average diameter between two measurement points of the voltmeter. An advantage of this method is that the resistance of clips has no influence on the results.

$$\rho = R \cdot \frac{A}{l_{\text{mag}}} = \frac{u}{i/2} \cdot \frac{A}{l_{\text{mag}}} \tag{3.1}$$

$$A = h \cdot (D_{\text{o}} - D_{\text{i}})/2 \tag{3.2}$$

$$l_{\text{mag}} = \frac{1}{6} \cdot \pi \cdot (D_{\text{o}} + D_{\text{i}})/2 \tag{3.3}$$

For each specimen, the above measurement has been carried out for many times, and the electrical resistivity are determined with the least square method to improve the accuracy. The results are listed in Table 3.2.

Specimen	**Resistivity** [Ωcm]
Somaloy 700HR3P	1.58E-2
Siron®S720	2.69E-2
Siron®S280b	16.8E-2
Siron®S300b	5.16E-2
Siron®S400b	8.00E-2
Siron®STestb	8.16E-2

Table 3.2. Electrical resistivity of different SMC

3.2.2 Commutation curve

The measurement of commutation curve (BH-curve) is carried out with both the commercial available system MPG200 of the firm Brockhaus Measurements and the self-developed measuring system. The results of both systems agree well with each other. Both systems have the same basic structure which is illustrated in Fig. 3.6.

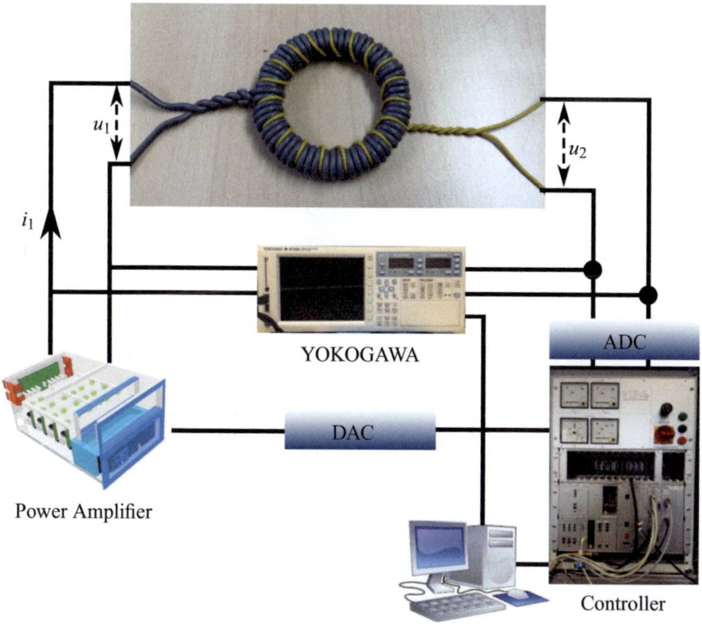

Figure 3.6. Measuring system with SMC ring specimen

The ring specimens are wound with two separate coils. The primary blue coil is fed with voltage u_1 which is generated with the help of a linear power amplifier regulated by the controller.

Due to the excited varying magnetic field in the ring specimen, voltage u_2 is induced in the secondary yellow coil. Assuming that the magnetic field in SMC ring is homogeneous, the commutation curve of SMC can be determined with the measured $i_1(t)$ and $u_2(t)$. Simultaneously, the iron losses are measured with a power analyzer WT3000 of the company YOKOGAWA.

According to the Ampère's circuital law, the magnetic field strength H can be indirectly calculated from current i_1 with (3.4) where N_1 is the number of turns of the primary coil.

$$\oint \vec{H}\,\mathrm{d}\vec{l} = N_1 \cdot i_1 \Longrightarrow H = \frac{N_1 \cdot i_1}{l_e} \tag{3.4}$$

The calculation of effective magnetic length of ring specimen l_e can be determined according to DIN EN 60205 using (3.5) where r_a and r_i are outer and inner radius of the SMC ring respectively.

$$l_e = 2 \cdot \pi \cdot \frac{\ln\left(\dfrac{r_a}{r_i}\right)}{\dfrac{1}{r_i} - \dfrac{1}{r_a}} \tag{3.5}$$

The magnetic flux density B in the ring specimen is determined from the induced voltage u_2, as shown in (3.6).

$$\begin{aligned} u_2 &= \frac{\mathrm{d}\Psi}{\mathrm{d}t} = N_2 \cdot \frac{\mathrm{d}\Phi}{\mathrm{d}t} = N_2 \cdot A_e \cdot \frac{\mathrm{d}B}{\mathrm{d}t} \\ B &= \frac{1}{N_2 \cdot A_e} \int_T u_2(t)\,\mathrm{d}t \end{aligned} \tag{3.6}$$

The effective cross sectional area A_e can be calculated using (3.7) where h is the thickness of SMC ring.

$$A_e = h \cdot \frac{\ln^2\left(\dfrac{r_a}{r_i}\right)}{\dfrac{1}{r_i} - \dfrac{1}{r_a}} \tag{3.7}$$

Because of hysteresis effect, the measured BH-curve is closed loop whose form depends on frequency and amplitude of the exciting current i_1. The commutation curve can be obtained by connecting the commutation points of all BH-loops. For instance, the BH-loops of SMC Siron®STestb for different amplitude of i_1 with the same frequency 50 Hz and the corresponding commutation curve are illustrated in Fig. 3.7.

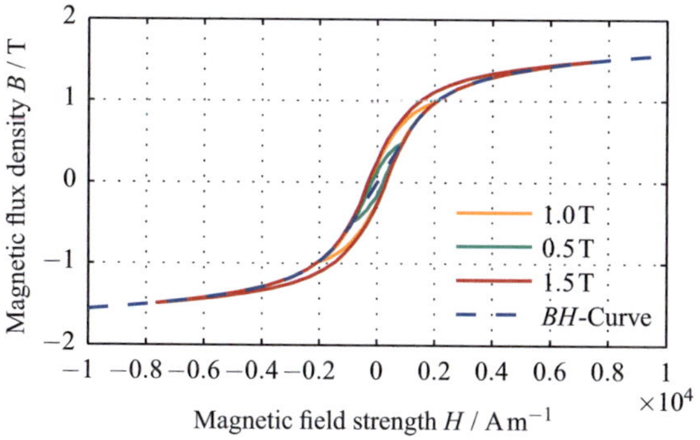

Figure 3.7. Definition of commutation curve

The commutation curves of SMC Siron®S280b corresponding to different frequencies are illustrated in Fig. 3.8. It can be noted the *BH*-curve of SMC indicating its magnetization ability remains almost unchanged at higher frequency. It implies that the induced eddy currents in SMC are very limited.

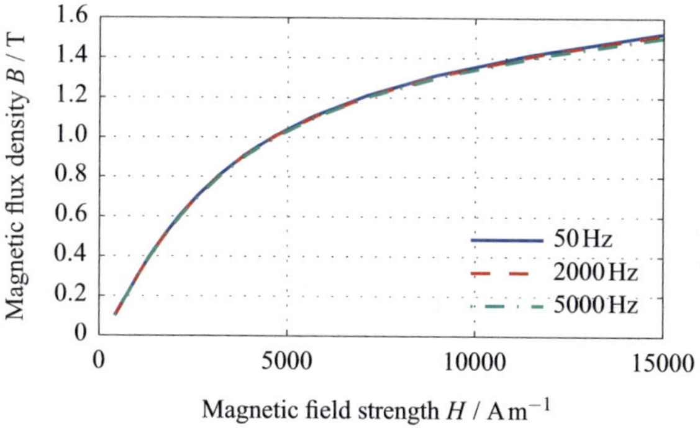

Figure 3.8. Commutation curves of Siron®S280b at different frequencies

In comparison, the *BH*-curves of laminated electrical steel M330-35A of the firm C.D. Wälzholz are illustrated in Fig. 3.9. The magnetization ability changes significantly depending on the frequency because of high eddy currents. It indicates that the working frequency of electrical steel is much more limited compared with SMC.

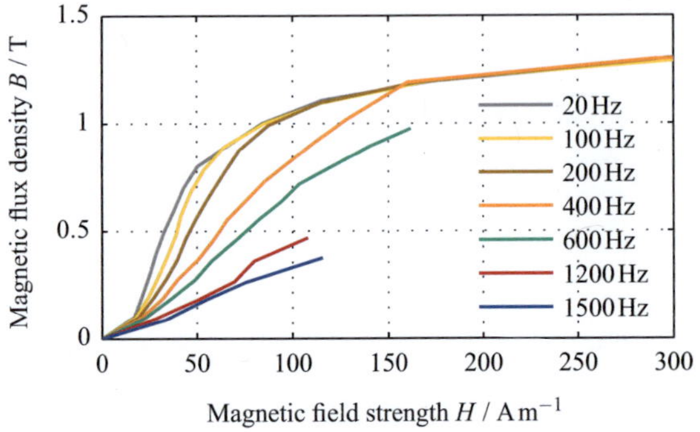

Figure 3.9. Commutation curves of NES M330-35A at different frequencies [241]

In general, only one *BH*-curve measured at low frequency is used in FEM and its dependence on the frequency can hardly be taken into account. As a result, the FEM calculation of electrical machine based on SMC is more accurate from this perspective. Based on the determined *BH*-curve, the relative permeability can be calculated using (3.8) where μ_0 is the vacuum permeability.

$$\mu_r = \frac{B}{\mu_0 \cdot H} \quad \text{with} \quad \mu_0 = 4\pi \cdot 10^{-7}\ \mathrm{H/m} \tag{3.8}$$

For instance, the commutation curve and the relative permeability of Somaloy 700HR3P are illustrated in Fig. 3.10. It should be noted the measured *BH*-curve should be modified before applied in FEM. First of all, the increase of permeability at the beginning of *BH*-curve leads to slow convergence or even divergence. Thus, these measuring points should be omitted.

On the other side, the maximum magnetic field strength is always restricted by the achievable current i_1 in the measurement. As a result, the *BH*-curve should be extrapolated to make sure that the B value is clearly defined for even much higher H in FEM.

One of the extrapolation methods is described as follows, where the μ_r, J_s and a are the initial relative permeability, saturation polarization and knee adjusting coefficient.

$$B = \mu_0 \cdot H + J_s \cdot \frac{H_a + 1 - \sqrt{(H_a + 1)^2 - 4H_a(1 - a)}}{2 \cdot (1 - a)} \quad (3.9)$$

$$H_a = \mu_0 \cdot H \cdot \frac{\mu_r - 1}{J_s} \quad (3.10)$$

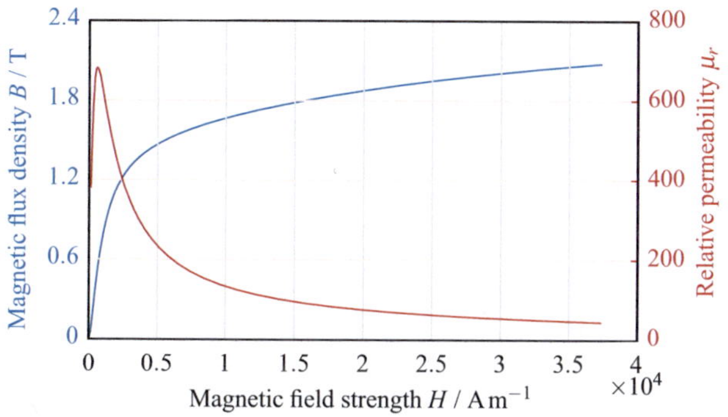

Figure 3.10. Commutation curve and permeability of Somaloy 700HR3P

The commutation curves of ring specimens in Table 3.1 for f equalling 50 Hz are illustrated in Fig. 3.11. For a better comparison, the commutation curves of the conventional laminated electrical steels NO20 and M235-35A are also added into Fig. 3.11.

There is no surprise that the non machinable SMC Somaloy 700HR3P and Siron®S720 have better performance because of the absence of organic binder. Surprisingly, the newly developed product Siron®STestb demonstrates relatively high permeability and saturation flux density compared with the other machinable SMC, even better than Siron®S720 at high magnetic field strength. In addition, it can be noted that the electrical steels have generally higher permeability than SMC. However, the difference becomes smaller with increasing H.

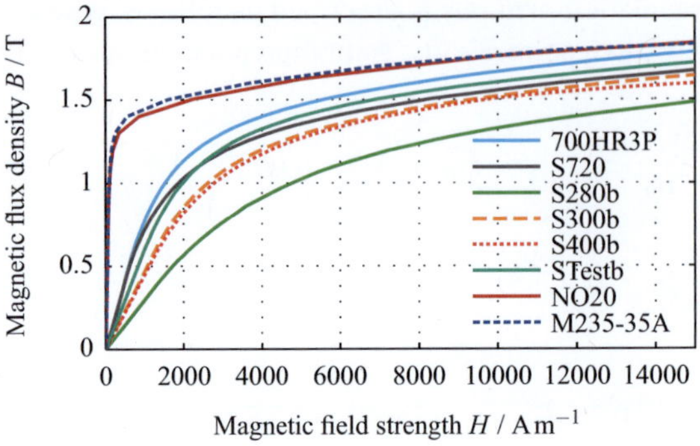

Figure 3.11. Commutation curves of SMC and NES

3.2.3 Iron losses

The iron loss is an important magnetic characteristic to evaluate SMC. Lower iron losses correspond to higher efficiency and less generated heat, which is of great importance for electrical machines.

With the help of power analyzer, the iron losses within the SMC ring are measured according to the norm DIN IEC 60404 – 6 for excitation of sinusoidal flux density with different frequencies and amplitudes. As a result, the induced voltage u_2 with the required sinusoidal form should be achieved through regulation u_1 where the geometric parameters of the ring, the number of turns of the coils and the required flux density should be taken into account.

The voltage of the primary winding is determined with (3.11).

$$u_1(t) = R_1 \cdot i_1(t) + u_{ind}(t) \tag{3.11}$$

where the u_{ind} is the induced voltage in the primary coil, which can be calculated with the induced voltage in the secondary coil:

$$u_{ind}(t) = u_2(t) \cdot \frac{N_1}{N_2} \tag{3.12}$$

Accordingly, the power of the primary winding is:

$$u_1(t) \cdot i_1(t) = R_1 \cdot i_1(t)^2 + u_2(t) \cdot \frac{N_1}{N_2} \cdot i_1(t) \tag{3.13}$$

Finally, the iron losses of the total ring specimen can be calculated with (3.14)

$$P_{fe} = \frac{1}{T} \cdot \int_T \frac{N_1}{N_2} \cdot u_2(t) \cdot i_1(t)\, \mathrm{d}t \tag{3.14}$$

In order to evaluate the deviation of u_2 from the perfect sinusoidal waveform, an form factor is defined with (3.15) in DIN IEC 60404 – 3 which equals 1.1107 in the ideal case.

$$F = \frac{U_{eff}}{|\overline{U}|} = \frac{\sqrt{\frac{1}{T} \int_T u_2^2(t)\, \mathrm{d}t}}{\frac{1}{T} \int_T |u_2(t)|\, \mathrm{d}t} \tag{3.15}$$

For instance, the form factors of all measurements of Siron®S300b are illustrated in Fig. 3.12. It can be seen that all the measurements have relatively small form factors and can be considered as valid. The largest deviation appears at high frequency 5000 Hz and high flux density.

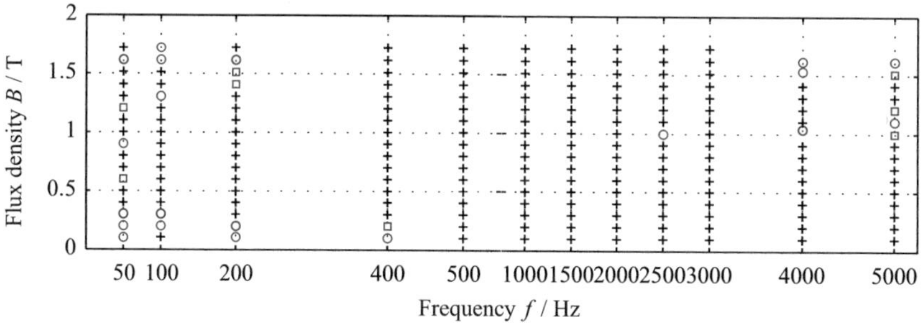

Figure 3.12. Form factors of measurements of Siron®S300b
+: deviation< 0.01 % ○: 0.05 % <deviation0.1 %
□: 0.01 % <deviation0.05 %

The iron losses of SMC rings are compared with that of three different laminated electrical steels of the company ThyssenKrupp in Fig. 3.13.

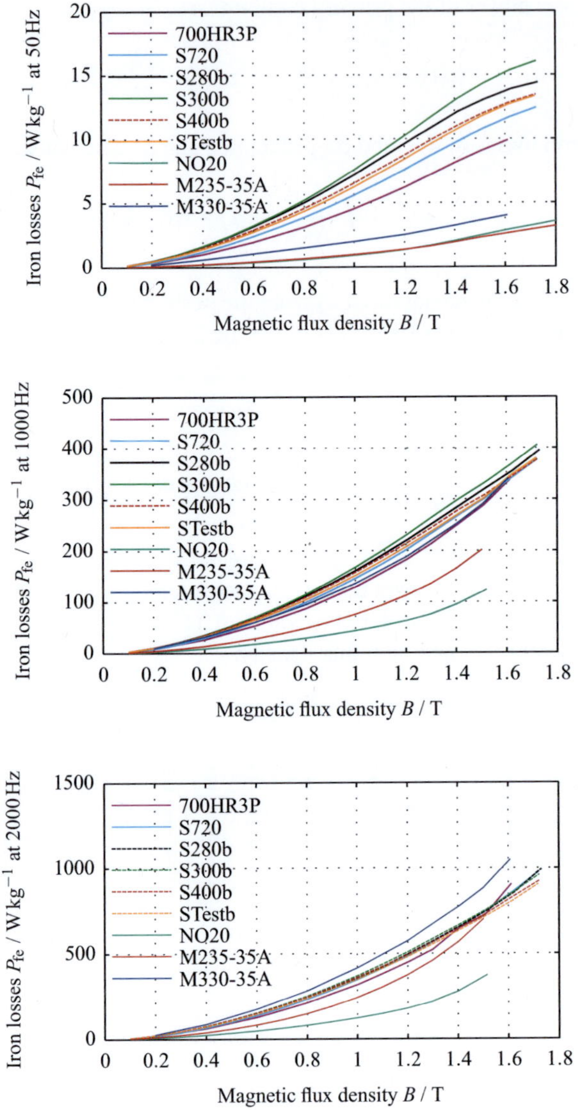

Figure 3.13. Iron losses of SMC and laminated electrical steels at different frequencies

It can be noted that the electrical steels have obviously lower iron losses than SMC at low frequency domain. However, the differences become smaller with increasing frequency.

At the frequency equal to 1000 Hz, the iron losses of M330-35A become comparable to Somaloy 700HR3P. The crossing between the M235-35A and SMC occurs at frequency 2000 Hz and induction 1.5 T. At this moment, the iron losses of M330-35A are already higher than SMC. It should also be noted that for the electrical steel NO20 with thickness 20 μm, the iron losses remain very low, even at extremely high frequency.

3.3 Iron loss calculation based on 3D FEM

The form and area of closed BH-loop vary with increasing frequency. Because the magnetic field in ring specimen is nearly homogeneous, the consumed energy during a complete magnetization process can be calculated as:

$$W_{fe} = V \cdot A_{loop} = V \cdot \oint H \, dB \tag{3.16}$$

V and A_{loop} are total volume of SMC ring and area of the BH-loop respectively.

Then, the specific iron losses in W/kg can be calculated with (3.17) where f is the frequency of magnetization.

$$P_{fe} = f \cdot \frac{W_{fe}}{\rho \cdot V} = f \cdot \frac{A_{loop}}{\rho} \tag{3.17}$$

Generally, the iron losses consist of hysteresis losses P_{hy} and dynamic losses P_{dyn}. The P_{hy} equals the area of the BH-loops for $f \rightarrow 0$. The dynamic losses P_{dyn} can be further divided into eddy current losses P_w and anomalous losses P_a.

$$P_{fe} = P_{hy} + P_{dyn} = P_{hy} + P_w + P_a \tag{3.18}$$

Accurate computation of iron losses is one of the most important and complicated tasks in the design and optimization of electrical machine As a consequence, great efforts have been made and several iron loss models have been suggested in the literatures.

The following model derived from classical Bertotti formula is utilized in this thesis to calculate the specific iron losses of sinusoidally excited SMC

$$P_{\mathrm{fe}} = k_{\mathrm{hy}}\, f\, \hat{B}^{(a\,\cdot\,\hat{B}^2 + b\,\cdot\,\hat{B} + c)} + k_{\mathrm{w}}\, f^2\, \hat{B}^2 + k_{\mathrm{a}}\, f^{1.5}\, \hat{B}^{1.5} \tag{3.19}$$

where f and $\hat{B}$ are frequency and amplitude of the sinusoidal magnetic field. The coefficients k_{hy}, a, b, c, k_{w} and k_{a} are determined based on the measured P_{fe} of different SMC materials with the Matlab surface fitting tool.

Because the iron loss calculation is carried out with the help of FEM, the measuring results should be converted into W/mm^3 before fitting. For instance, the measured P_{fe} and the fitted iron loss curves of Somaloy 700HR3P are illustrated in Fig. 3.14.

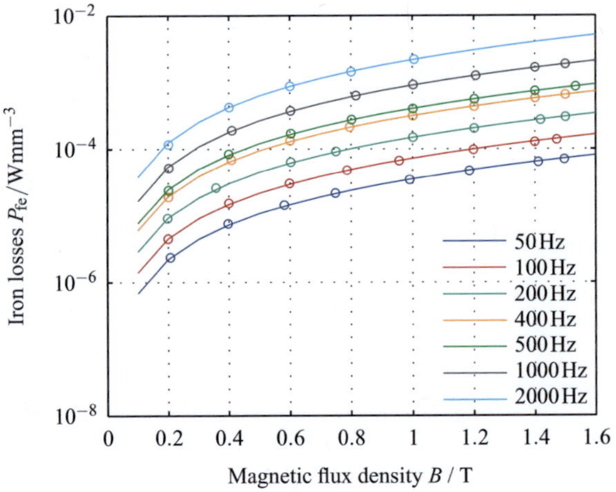

Figure 3.14. Fitting of the measured iron losses of Somaloy 700HR3P

It should be noted that the model described in (3.19) is only valid for the alternating and sinusoidal magnetic field. However, many harmonics exist in electrical machine and real B field in SMC is far away from sinusoidal.

The harmonics contained in the flux density waveform are determined with Fourier analysis and the iron loss model can be improved as

$$P_{\mathrm{fe}} = k_{\mathrm{hy}}\, f \hat{B}_1^{(a \cdot \hat{B}_1^2 + b \cdot \hat{B}_1 + c)} + \sum_i \left(k_{\mathrm{w}}\, (i \cdot f)^2\, \hat{B}_i^2 + k_{\mathrm{a}}\, (i \cdot f)^{1.5}\, \hat{B}_i^{1.5} \right) \tag{3.20}$$

where $f, \hat{B}_1$ are frequency and amplitude of the fundamental harmonic while $\hat{B}_i$ is the amplitude of the ith harmonic contained in flux density waveform.

The complete model is discretized into a large number of small elements in FEM and there are three components forming the B field in each SMC element. Thus, the final form of iron loss model should be

$$\begin{aligned} P_{\mathrm{fe}} = & \sum_j k_{\mathrm{hy}}\, f \mid \hat{B}_1 \mid^{(a \cdot |\widehat{B}_1|^2 + b \cdot |\widehat{B}_1| + c)} + \\ & \sum_j \left(\sum_i \left(k_{\mathrm{w}}\, (i \cdot f)^2\, |\widehat{B}_i|^2 + k_{\mathrm{a}}\, (i \cdot f)^{1.5}\, |\widehat{B}_i|^{1.5} \right) \right) \end{aligned} \tag{3.21}$$

where j is the number of SMC element in FEM model and $|\widehat{B}_i|$ is the resultant magnitude of the ith harmonic contained in flux density waveform. It can be calculated with the corresponding components $\widehat{B}_{xi}$, $\widehat{B}_{yi}$, and $\widehat{B}_{zi}$ along the x, y, and z axes respectively.

$$|\widehat{B}_i| = \sqrt{\hat{B}_{xi}^2 + \hat{B}_{yi}^2 + \hat{B}_{zi}^2} \tag{3.22}$$

The $|\widehat{B}_i|$ can be calculated using the three components $\widehat{B}_{ri}$, $\widehat{B}_{\theta i}$, $\widehat{B}_{zi}$ as well. They correspond to the peak values of the ith harmonic contained in the flux density waveforms along the radial, tangential and z directions respectively.

$$|\widehat{B}_i| = \sqrt{\hat{B}_{ri}^2 + \hat{B}_{\theta i}^2 + \hat{B}_{zi}^2} \tag{3.23}$$

3.4 Validation of the iron loss model

The accurate calculation of iron losses is very important for the performance evaluating of electrical machine [S1]. Therefore, a test system consisting of a yoke and a tooth has been built to validate the developed iron loss model, as illustrated in Fig. 3.15.

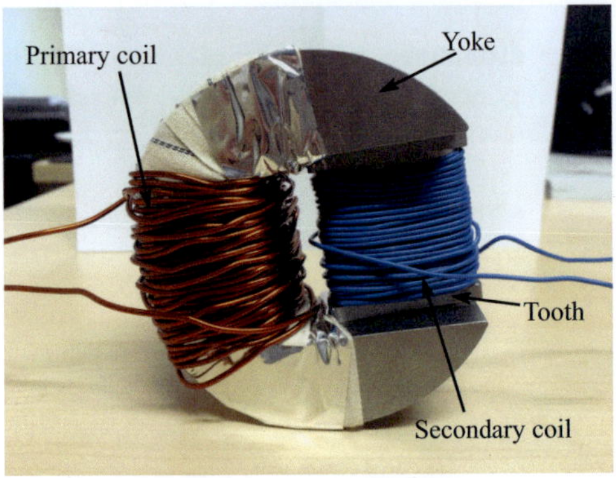

Figure 3.15. Test system to validate the iron loss model

Both components are made from the same material Siron®STestb. The primary coil wound on the yoke is excited with sinusoidal voltage u_1 whose frequency and amplitude can be regulated with a controller. Meanwhile, the iron losses are measured with a power analyser.

On the other hand, the test system is modelled and analysed with the help of 3D FEM. For instance, the meshed model and the results including the *B*-field at a specific time instant are illustrated in Fig. 3.16 where the small red arrows indicate the direction of magnetic flux.

The electrical circuit of FEM model is illustrated in Fig. 3.17. In order to eliminate the inaccuracy resulted from the measured resistance of the primary coil, five current sources instead of a voltage source are connected in parallel to feed the primary coil whose parameters are determined from the Fourier analysis of the measured current. The R_1 represents the resistance of the primary coil which however has no influence on magnetic field and iron losses. The R_2 is a very large resistance to approximate open circuit state of the secondary coil.

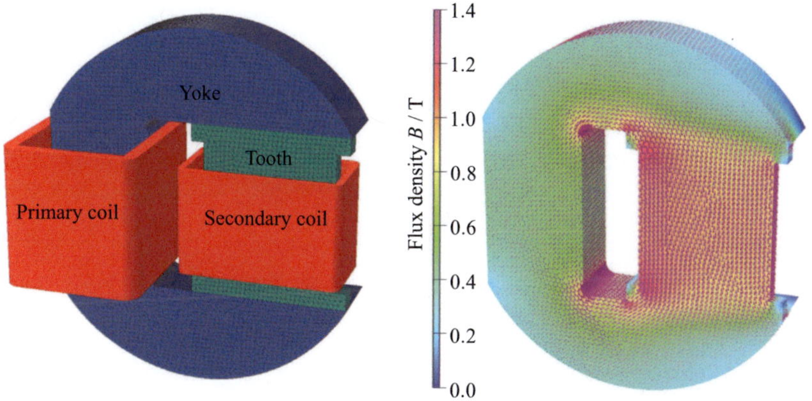

(a) 3D FEM model (b) B-Field and magnetic flux

Figure 3.16. 3D FEM model and the FEM results

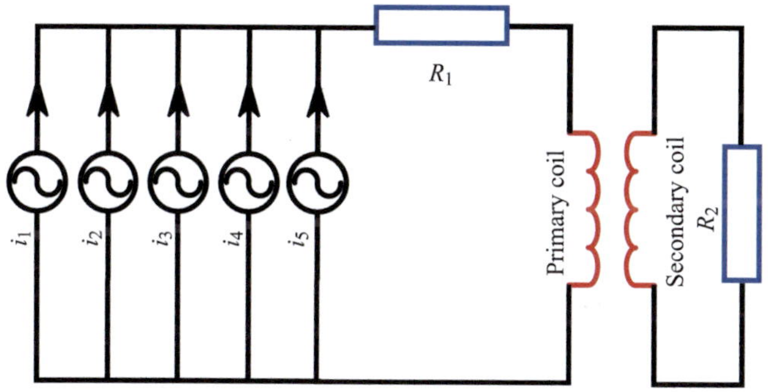

Figure 3.17. Electrical circuit of the 3D FEM model

The iron losses of yoke and tooth are calculated based on 3D FEM using (3.21) for different voltage u_1 which are listed together with the measuring results in Table 3.3.

f	$P_{fe}(24\,V)$			$P_{fe}(32\,V)$			$P_{fe}(36\,V)$		
[Hz]	**Mes.**	**FEM**	**Error**	**Mes.**	**FEM**	**Error**	**Mes.**	**FEM**	**Error**
100	16.7 W	16.3 W	2.4%	24.3 W	24.6 W	1.1%	27.3 W	29.3 W	7.4%
200	12.0 W	11.6 W	3.3%	19.3 W	18.4 W	4.8%	23.8 W	22.2 W	7.0%
400	8.31 W	8.12 W	2.2%	13.7 W	12.8 W	7.0%	16.5 W	15.8 W	4.3%
500	7.34 W	7.44 W	1.3%	12.3 W	11.5 W	6.2%	14.7 W	14.3 W	3.0%
1000	4.91 W	6.08 W	23.8%	8.49 W	9.40 W	10.7%	10.2 W	11.5 W	13.5%

Table 3.3. Comparison of the measured and calculated iron losses

It can be noted that the deviation is lower than 8% for the u_1 with frequency lower than 500 Hz which increases with higher frequency, especially for the u_1 with low amplitude. The most important reasons leading to the deviation are:

Existence of rotational field

The parameters of iron loss model are determined from the measurement of ring specimens under alternating magnetic field. However, there isn't only alternating but also rotational magnetic field existing in the test system. For instance, the magnetic field in a SMC element of yoke is illustrated in Fig. 3.18 where the flux density B_z is very small and neglected for a better clarity.

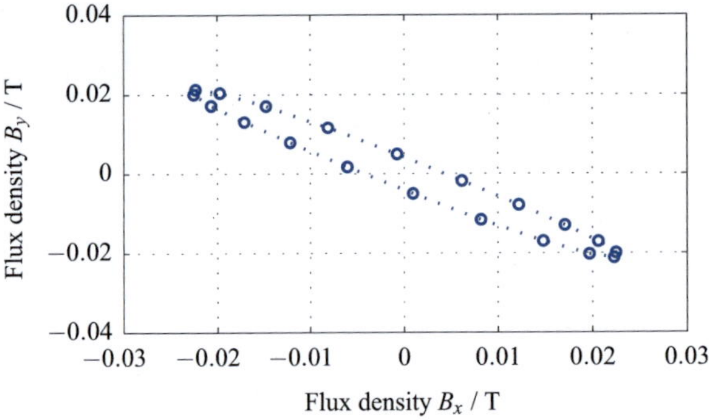

Figure 3.18. The magnetic field of a SMC element in yoke

A more accurate iron loss model can be developed only if the iron losses of SMC specimen under rotational magnetic field are available [233, 242, 243].

Influence of the machining

As stated above, the SMC including binder material can be machined after compaction and heat treatment. However, the insulation layer can be destroyed during machining and the magnetic properties of SMC components might be deteriorated, as illustrated in Fig. 3.19. Because of the damage of the insulation layer around the particles, the eddy current losses might become larger, especially for high frequency.

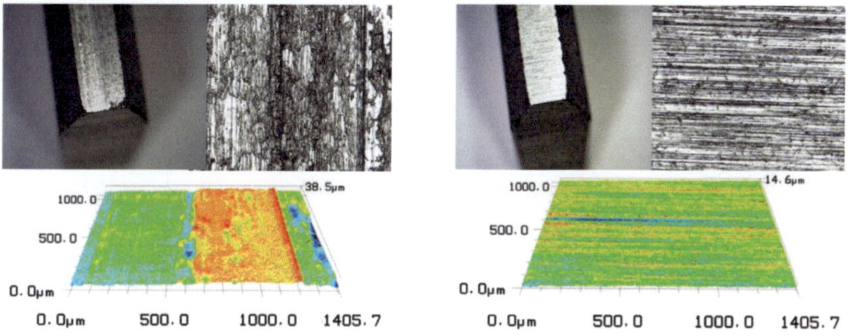

(a) Milling: roughness $\approx 30\,\mu m$ (b) Grinding: roughness $\approx 15\,\mu m$

Figure 3.19. Surface of SMC parts after machining [244]

Measuring error of current in primary coil

The parameters of current sources in the electrical circuit illustrated in Fig. 3.17 are calculated from i_1 whose measurement is based on the Ohm's law. As a result, the unavoidable measuring error of i_1 has great influence on the results, especially for the low value of i_1.

Error of mathematical fitting

The coefficients of iron loss model are determined based on the measured iron losses of SMC rings for different magnetic flux density and frequency through curve fitting. The errors resulted from the mathematical fitting are inevitable and can influence the calculated iron losses.

Air gap between yoke and tooth

Both the yoke and tooth are milled from standard cylinder. Between the two components, there are small air gaps which cannot be accurately measured and modelled in the FEM. However, it has large influence on the iron losses because the magnetic fields in both components can be significantly reduced due to the large reluctance of air.

Existence of leakage flux

The iron loss of the test system is determined with the measured i_1 and u_2. Compared with the measurement with ring specimen, the yoke and tooth have different cross sectional area. For this reason, the leakage flux in the test system is considerable, especially when high magnetic saturation exists.

In spite of above errors, the measurement results match relatively well with the calculated iron losses based on 3D FEM. Furthermore, the model might be more accurate for the iron loss calculation of electrical machine because the small gap between the yoke and tooth doesn't exist. In addition, the rotational magnetic field exists only in small regions of electrical machine. In conclusion, the developed model is valid and can be utilized for the iron loss calculation of electrical machine.

4

Development of a Simulation Platform based on 3D FEM

In order to analyse and optimize different electrical machines efficiently, a simulation platform based on 3D FEM has been developed. Owing to the script-based modelling, solving and post-processing, the analysis of a large number of electrical machines with different geometries, materials, excitations and so on becomes possible [O1].

For a specific stator of PMAFM, many different layouts of double layer concentrated winding (DLCW) are suitable. In order to compare diverse layouts and to select the most suitable DLCW, a winding editor is necessary. With its help, the most important characteristics of DLCW can be determined.

Another important improvement of the simulation platform is the high efficient analysis of PMAFM. On this basis, it is only necessary to perform FEA for one-sixth instead of an entire electrical period to determine the performance of a PMAFM. Hereby, the greatest challenge is the iron loss calculation based on the limited FEM results. By this means, computational time is significantly reduced, especially for the time-consuming 3D FEM.

In this chapter, basic structure of the simulation platform is firstly briefly described. Subsequently, the most important challenges of the simulation platform are presented which include the analysis of DLCW with a winding editor and the high efficient iron loss calculation of PMAFM.

4.1 Structure of the simulation platform

The 3D simulation platform comprises mainly three basic parts, namely pre-processor, solver and post-processor. Its basic structure is illustrated in Fig. 4.1. The complete process is based on many scripts so that only the essential parameters of the selected electrical machine need to be specified manually. Afterwards, the modelling, solving and analysis are carried out without intervention of user.

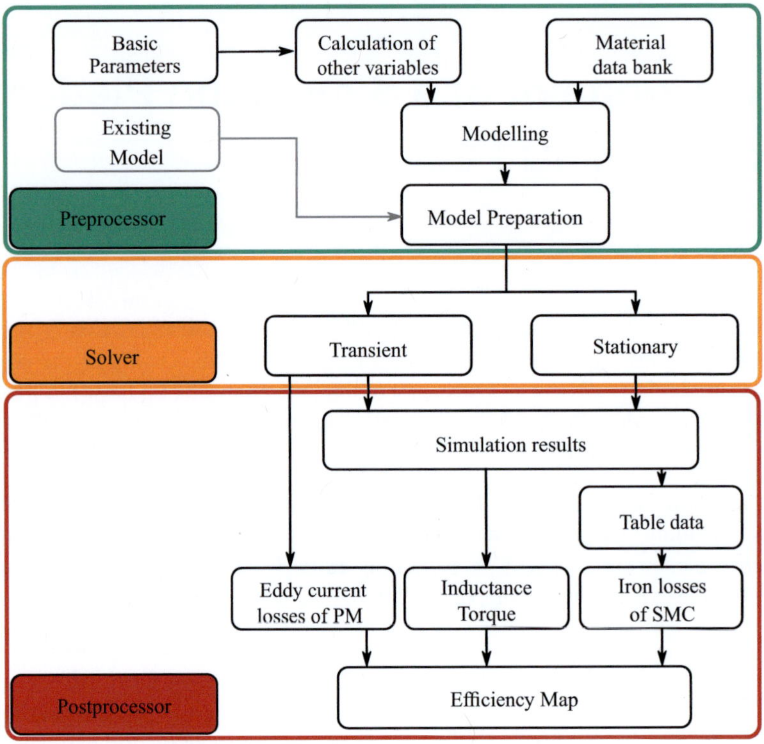

Figure 4.1. Structure of the 3D Simulation Platform

Preprocessor

The basic parameters can be divided into four categories: the geometric parameters, the parameters of materials including PM and SMC, the specifications of rotor movement and the parameters describing winding excitation.

With the help of geometric parameters and iron loss model described in chapter 3, the other necessary parameters can be calculated. Subsequently, modelling of the electrical machine can be carried out automatically and the model should be further prepared before solving. For instance, the complete FEM model should be discretized into a large number of small elements. In addition, the specifications of material properties, rotor movement and electrical excitation should be performed.

Solver

The Maxwell's equations are the basis to determine electromagnetic field in an electrical machine. Hereby, the displacement current is normally neglected due to the relatively low operating frequency. The $\vec{E}$, ρ and ε_0 are the electric field intensity, electric charge density and permittivity of free space respectively.

$$\begin{aligned}
\nabla \times \vec{H} &= \vec{J} \\
\nabla \times \vec{E} &= -\frac{\partial \vec{B}}{\partial t} \\
\nabla \cdot \vec{B} &= 0 \\
\nabla \cdot \vec{E} &= -\frac{\rho}{\varepsilon_0}
\end{aligned}$$

In general, the current density J consists of three components: the current density externally injected into the coil $\vec{J}_e$, the eddy current density in conductive material induced by time-varying magnetic field and the current caused by relative movement between magnetic field and conductive material. In (4.1), σ is the electrical conductivity, while $\vec{v}$ is the relative velocity between conductive material and mgnetic field.

$$\vec{J} = \vec{J}_e + \sigma \cdot \vec{E} + \sigma \cdot \vec{v} \times \vec{B} \tag{4.1}$$

The prepared model can be analysed either transiently or stationary. The transient analysis is more time-consuming than stationary analysis. However, it is more accurate because all time-varying effects can be taken into account. For instance, the eddy currents induced in PMs during the rotation can only be determined with the help of transient analysis.

Because the divergence of B equals 0, the magnetic flux density B can be defined as rotation of a vector potential $\vec{A}$ whose divergence equals 0. The following formulation is used in transient analysis.

$$\vec{B} = \nabla \times \vec{A} \quad \text{with} \quad \nabla \cdot \vec{A} = 0 \tag{4.2}$$

For the case of 2D FEA, the vector potential $\vec{A}$ has only one component $\vec{A}_z$ to be solved while the components $\vec{A}_x$ and $\vec{A}_y$ equal 0. As a result, the calculation complexity and computation time are significantly reduced. In comparison, all three components of $\vec{A}$ need to be determined in 3D FEA.

For the stationary analysis, the time-varying parameters and effects aren't taken into account [O2, S2]. For this circumstance, the current density is assumed to be constant and equals the externally injected current density.

$$\vec{J} = \vec{J}_{\mathrm{e}} \tag{4.3}$$

In order to reduce the computation time for stationary analysis with 3D FEA, the total FEM model is split into the part V_{s} where the injected current flows and the rest part V_{r}. Within the part V_{s}, the total magnetic field strength $\vec{H}$ consists of reduced field strength $\vec{H}_{\mathrm{m}}$ and conductor field strength $\vec{H}_{\mathrm{s}}$.

$$\vec{H} = \vec{H}_{\mathrm{m}} + \vec{H}_{\mathrm{s}} \tag{4.4}$$

The $\vec{H}_{\mathrm{s}}$ represents the magnetic field intensity generated at position r by the injected steady current at position r' which can be calculated with the Biot-Savart law. Hereby, V is the volume of Biot-Savart conductor.

$$\vec{H}_{\mathrm{s}} = \frac{1}{\mu} \cdot \frac{\mu_0}{4\pi} \int_V \frac{\vec{J} \times (\vec{r} - \vec{r'})}{|\vec{r} - \vec{r'}|^3} \, \mathrm{d}V \tag{4.5}$$

In comparison, the magnetic field strength $\vec{H}_\mathrm{m}$ is generated by magnetized parts such as PM. It can be represented as gradient of a scalar potential φ.

$$\vec{H}_\mathrm{m} = -\nabla \cdot \varphi \tag{4.6}$$

In the rest part V_r, the total magnetic field strength $\vec{H}$ is equal to $\vec{H}_\mathrm{m}$ due to the absence of injected current. On the interface of two different types of parts, the scalar potential can be coupled by applying the continuity condition.

Compared with the vector potential formulation where three components of $\vec{A}$ need to be solved for 3D FEA, only one component φ needs to be determined in scalar potential formulation, which leads to considerable reduction of computational time. Besides, the FEA for stationary analysis is more robust because the whole model should be newly meshed for every rotor position. In comparison, the model for transient analysis is only meshed at initial position. Afterwards, only a part of the air gap is remeshed. However, the transient analysis is more accuracy because all time-varying effects such as induced eddy currents should be taken into account, which implies the convergence is much slower.

Postprocessor

Based on the solved electromagnetic field, the electrical machine can be fully characterized in post-process. The results can be divided into two groups. Some of them can be directly exported from FEA such as torque. The rest are the so-called table data where the flux density components and coordinates of each SMC element in the FEM model are saved. Based on the table data, the iron losses in SMC can be calculated. As long as all the results are determined, the most informative efficiency map can be calculated.

4.2 Analysis of double layer concentrated winding

The PMAFM is a promising topology for the application of SMC due to its 3D magnetic flux and complex cores. The DLCW illustrated in Fig. 4.2 is frequently utilized in PMAFM because of many distinguished advantages [245–250].

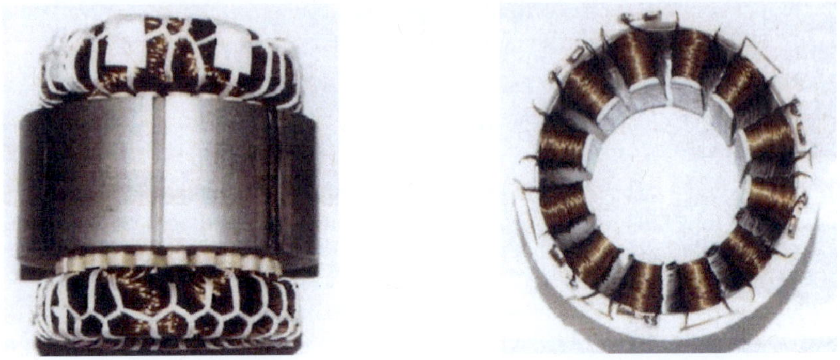

(a) Distributed winding (b) Concentrated winding

Figure 4.2. Comparison of distributed and concentrated winding [245]

Compared with distributed winding, concentrated winding has simpler structure and shorter end-windings, which leads to reduction of resistance and copper losses. In addition, the phases are physically isolated from each other because there is no overlapping among them, which can effectively improve the fault tolerance. Furthermore, a high filling factor can be achieved and the generated heat can be more effectively transmitted to the stator. Finally, the automated manufacturing of concentrated winding can be more easily realized, which results in reduction of costs for mass production. However, much more harmonics are contained in the magnetomotive force (MMF) compared to distributed winding, which can cause higher torque ripple, noise and eddy current losses in the electrically conductive material.

In order to analyse different layouts of DLCW, a winding editor based on analytical calculation has been developed to provide indications in the design process. The winding editor is integrated in the simulation platform as well [S3]. Its graphical user interface (GUI) is illustrated in Fig. 4.3.

First of all, a definition to describe the layout of DLCW is presented. Subsequently, calculation of the most important properties such as winding factor and MMF harmonic contents are presented.

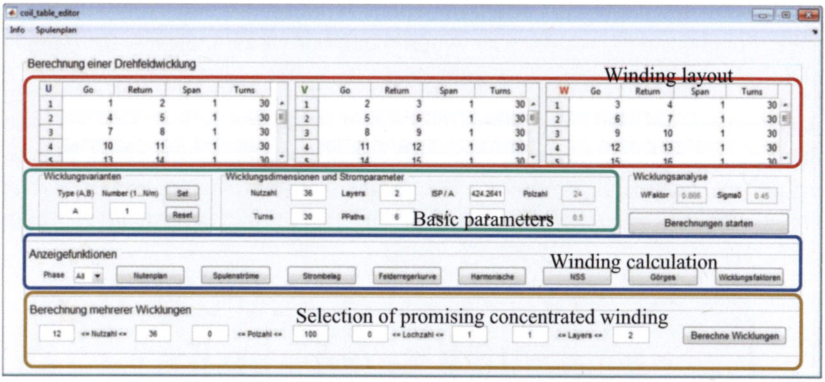

Figure 4.3. GUI of the winding editor for DLCW analysis

4.2.1 Classification of concentrated winding

For a stator with determined number of slots, many symmetrical concentrated windings with different layouts might be suitable. In order to clearly describe the DLCW, the following definition is introduced:

$$
U \times M\ V\ N \quad
\begin{cases}
U\text{: Number of elementary winding} \\
M\text{: Number of layers of concentrated winding} \\
V\text{: Winding variant} \\
N\text{: Winding number}
\end{cases}
$$

Elementary winding is the basic unit of DLCW. The complete winding can be obtained by repeating the elementary winding sequentially U times. The concentrated winding can be realized with either one or double layers in each slot of stator. In comparison, the DLCW has shorter end-windings and more sinusoidal back electromotive force (emf) than one layer concentrated winding. Therefore, only the DLCW is used in this thesis, which implies M equals 2. Basically, there are two possible winding variants. For the winding variant A, there is only one coil group of each phase in an elementary winding, and the current direction of different phases are identical. For the winding variant B, there are two coil groups of each phase in an elementary winding. Both coil groups have inverse current directions and the current directions of adjacent phases are inverted as well. Finally, the N is defined as winding number which indicates the number of adjacent coils belonging to a coil group.

According to the above definition, the DLCW illustrated in Fig. 4.4 can be defined as $1 \times 2B2$. There are 12 slots in the stator, and the utilized DLCW comprises only 1 elementary winding. In addition, there are two coil groups of each phase with inverse current directions, and each coil group is made up of 2 adjacent coils.

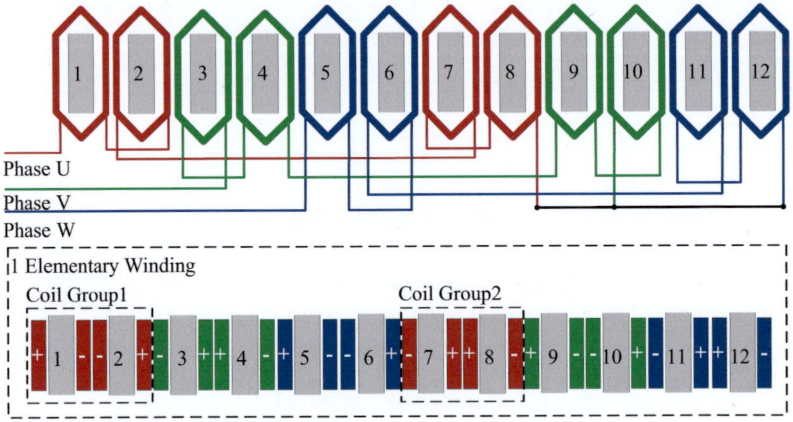

Figure 4.4. Winding layout of DLCW $1 \times 2B2$

Another two possible winding designs suitable for the same stator are $4 \times 2A1$ and $2 \times 2A2$, whose layouts are shown in Fig. 4.5.

It can be noted that for the winding variant A the current directions of coil groups of different phases are the same and there is only one coil group belonging to each phase in an elementary winding.

For the DLCW $4 \times 2A1$, there are totally 4 elementary windings and each coil group comprises only one coil. While for the DLCW $2 \times 2A2$, there are 2 elementary windings and each coil group is made up of two adjacent coils.

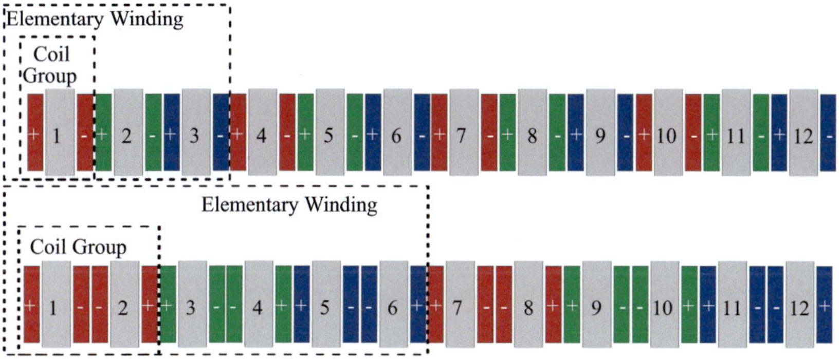

Figure 4.5. Winding layout of DLCW 4 × 2A1 and 2 × 2A2

4.2.2 Analysis of DLCW

As long as the winding layout of DLCW is determined, the most important characteristics of DLCW can be calculated analytically based on the following simplifying assumptions:

- Permeability of the soft magnetic material is infinitely large so that the magnetic flux travels vertically into the stator and the rotor.
- Magnetic flux in the air gap comprises only radial component for RFM and only axial component for AFM respectively. Compared with the magnetic flux in air gap, the end leakage flux is very small and can be neglected.
- Slot opening is very small compared with the air gap length so that the MMF generated by conductors in a slot can be considered as impulse at slot opening.

There are many different methods to analyse the DLCW and the complete analysing process of DLCW in the developed coil editor is illustrated in Fig. 4.6.

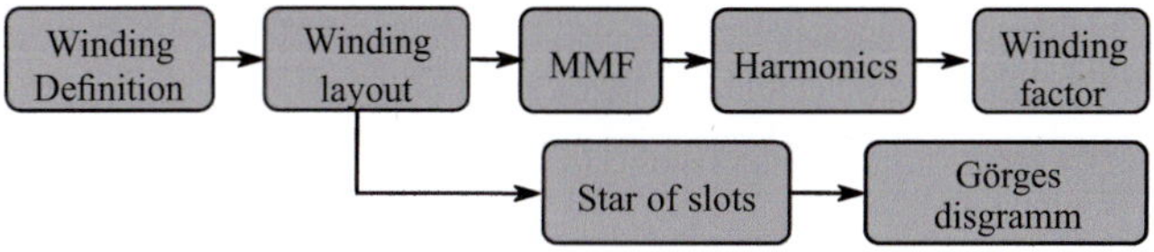

Figure 4.6. Analysis process of DLCW

Subsequently, the complete analysing process is explained for the DLCW $1 \times 2B2$ shown in the Fig. 4.4. It should be noted the two coil groups of each phase are connected in parallel. The sinusoidal current in the three phases are defined as follows, where $\hat{I}$ is the amplitude of the phase current.

$$\begin{aligned} i_U &= \hat{I} \cdot \cos(wt) \\ i_V &= \hat{I} \cdot \cos(wt - \frac{2}{3}\pi) \\ i_W &= \hat{I} \cdot \cos(wt - \frac{4}{3}\pi) \end{aligned}$$

Assuming that $\hat{I} = 10$ A and the number of turns of coil around one stator tooth of one phase equals 20, the total current concentrated as impulse at slot opening at the time instant $t = 0$ can be determined. On this basis, the MMF generated by the current of all these three phases can be calculated with the Ampère's circuital law. The MMF harmonic content can subsequently be determined with the Fourier analysis, as shown in Fig. 4.7.

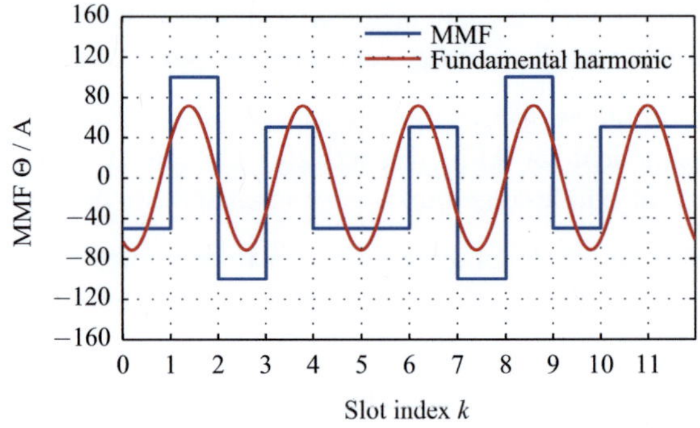

Figure 4.7. MMF and the fundamental harmonic of DLCW $1 \times 2B2$

The harmonics contained in MMF are of great importance as they are the main cause of the rotor losses. A comparison of the amplitude of the harmonics can be found in Fig. 4.8. The order of harmonics are equal to 1, 5, 7, 11, 13, 17, 19 and so on, while the fifth harmonic has the highest amplitude and works as fundamental harmonic.

It should be noted only the fundamental harmonic runs synchronously with rotor, while the other harmonics are unwanted and asynchronous. Especially the subharmonics can cause significant alternating magnetic fields in rotor, which leads to higher iron losses and higher eddy current losses in PMs [247, 251].

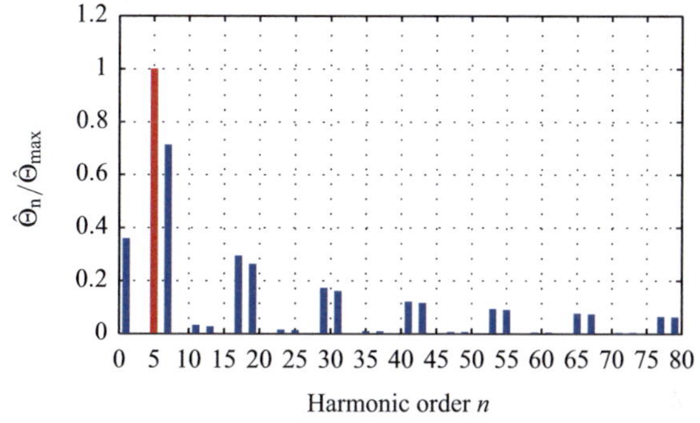

Figure 4.8. Harmonics contained in MMF of DLCW 1 × 2B2

Another important criterion to evaluate DLCW is the winding factor, which indicates the utilization rate of the total current in torque generation. A higher winding factor of the fundamental harmonic implies that higher effective voltage can be induced in the winding at the same rotational speed for the same magnetic flux generated by PM. The winding factor is equal to the product of distribution factor k_d, pitch factor k_p and skewing factor k_{skew}.

$$k_w = k_d \cdot k_p \cdot k_{skew} \tag{4.7}$$

The analytical calculation of winding factor has been suggested in many literatures. For most methods, great attentions should be paid to examine if the required preconditions are fulfilled. In this thesis, a general method based on the results of Fourier analysis of MMF is utilized, which takes both the distribution and pitch factor into account [252].

$$k_{w,n} = \frac{c_n \cdot n \cdot \pi}{4 \cdot \sum_{q=1}^{N_s} (N \cdot i)_q} \tag{4.8}$$

The c_n is the amplitude of the n-th harmonic in the MMF and $\sum_{i=1}^{N_s}(N \cdot i)_q$ is equal to the sum of the ampere-turns of all the N_s coils.

The calculated winding factors of different harmonics corresponding to the DLCW 1 × 2B2 are illustrated in Fig. 4.9.

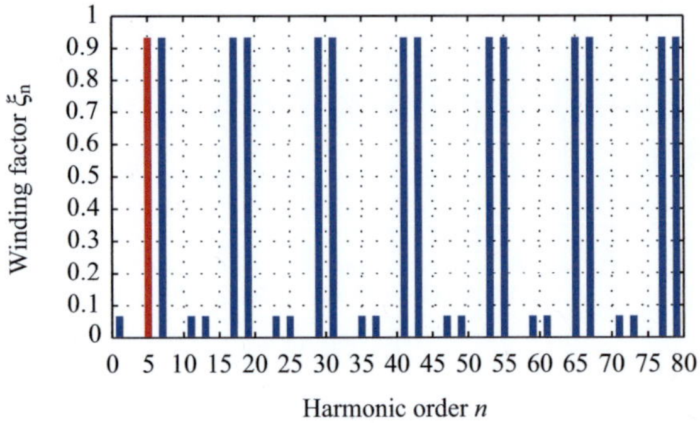

Figure 4.9. Winding factors of DLCW 1 × 2B2

In order to evaluate the harmonic content of different DLCW, the leakage factor σ_d is defined in the thesis. A lower leakage factor implies lower harmonic content and therefore lower losses.

$$\sigma_d = \frac{\sum_{n \neq p}^{\infty} \left(\frac{k_{w,n}}{n} \right)^2}{\left(\frac{k_{w,p}}{n} \right)^2} \tag{4.9}$$

A simple and efficient method to examine the symmetry of the DLCW is based on the Görges diagram, which is derived from star of slots. It is made up of the EMF phasor of each winding element in all slots. For instance, the star of slots and the Görges diagram of the DLCW 1 × 2B2 are illustrated in Fig. 4.10. The resultant EMF phasor of each phase is the vectorial summation of all the winding element phasor belonging to each phase.

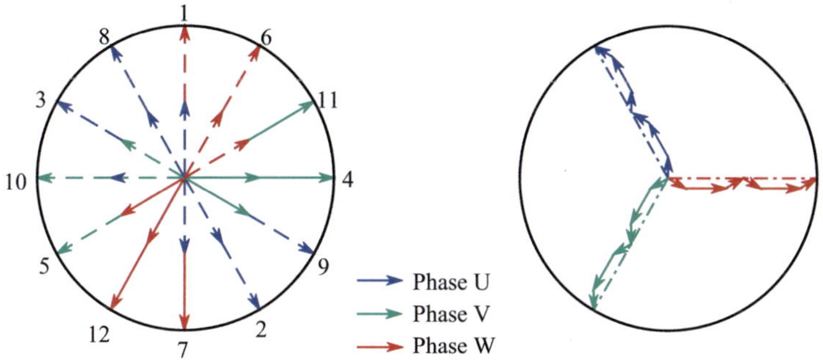

Figure 4.10. Star of slots and Görges diagram of DLCW 1 × 2B2

The winding factor of the fundamental harmonic can also be calculated as the ratio between the resultant phase EMF phasor and the scalar summation of all winding element phasor. As a result, the winding element phasors should be aligned as well as possible to reach a better utilization of the current.

4.3 Efficient calculation of iron losses of PMAFM

As mentioned above, the 3D FEM is necessary to analyse the PMAFM accurately. Because the 3D FEM is so time-consuming, the electromagnetic symmetry and periodicity should be fully utilized to reduce the computation time. In general, it is only necessary to simulate the PMAFM for one-sixth electrical period in order to completely characterize its performance, which is evident through the observation of the torque and flux linkage of PMAFM during the rotation.

However, the iron loss model defined in (3.21) requires the peak value of the harmonics contained in the flux density waveform (FDW) of each SMC element. Hence, the complete FDW of each SMC element for a whole electrical period should be reconstructed based on the limited FEA over $1/6$ electrical period [O3–O5].

In this section, the electromagnetic periodicity of PMAFM is firstly investigated through observing of the FEM results. Subsequently, the reconstruction of the FDW for the SMC elements in the stator and the rotor are explained respectively. Although the research in this chapter is carried out for the PMAFM-ISA, the method can be applied to iron losses calculation of other PMAFM or even PMRFM after slight revise.

4.3.1 Electromagnetic periodicity in PMAFM-ISA

The electromagnetic periodicity is investigated with a PMAFM-ISA with the DLCW $12 \times 2A1$, which consists of 36 stator segments and 24 poles. Considering the electromagnetic symmetry around the rotational axis and the flux normal boundary condition applied on the middle plane of the FEA model, it is only necessary to analyse $1/24$ of the entire model. The complete PMAFM-ISA and the 3D FEM model are illustrated in Fig. 4.11.

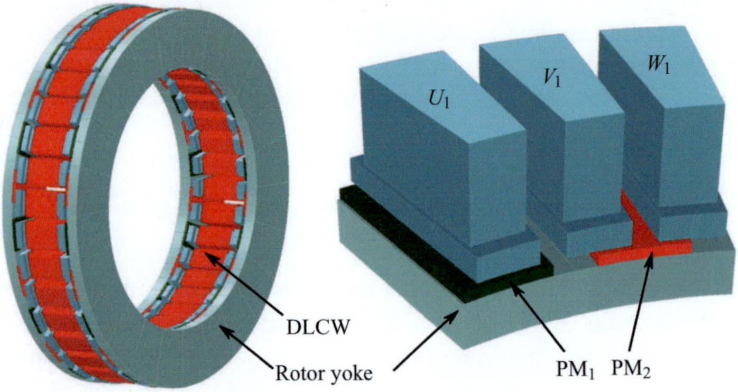

Figure 4.11. 3D FEM model of a PMAFM-ISA with DLCW $12 \times 2A1$

For the PMAFM-ISA, the currents of the three phases are defined as follows. For the initial rotor position $(t = 0)$, a PM is aligned with the stator segment U_1.

$$i_U = \hat{I} \cdot \cos(wt + 90^\circ)$$
$$i_V = \hat{I} \cdot \cos(wt - 30^\circ)$$
$$i_W = \hat{I} \cdot \cos(wt - 150^\circ)$$

The electromagnetic periodicity can be investigated with help of the FEA results. The flux density component B_z of the PMAFM-ISA for the rotor positions 0° and 10° are illustrated in Fig. 4.12. As the number of pole pairs of the PMAFM-

ISA is 12, the rotor positions correspond to time instants t equalling to 0 s and 1/3T respectively.

It can be seen that the B_z of the stator segment U_1 for the rotor position 10° is the same as that of the stator segment V_1 for the rotor position 0°. The reason is that the relative positions between the stator segments and the PM are identical for both positions. For the initial rotor position 0° ($t = 0$T), the PM_1 is aligned with the stator segment U_1 and PM_2 is aligned with the slot between segments V_1 and W_1. While for the rotor position 10° ($t = 1/3$T), the PM_1 is aligned with the stator segment V_1 and PM_2 is aligned with the slot between the U_1 and W_1.

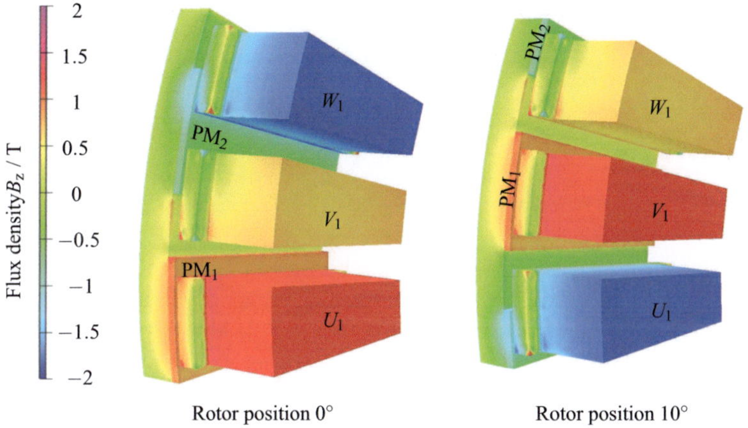

Figure 4.12. Flux density B_z of the PMAFM-ISA at the rotor positions 0° and 10°

On the other hand, the current flowing in the coils plays a dominant role in determining the magnetic field as well. For the time instant t equalling to 1/3T, the currents in the three phases are calculated as follows.

$$i_U(t = 1/3T) = \hat{I} \cdot \cos(210°) = i_W(t = 0\,s)$$
$$i_V(t = 1/3T) = \hat{I} \cdot \cos(90°) = i_U(t = 0\,s)$$
$$i_W(t = 1/3T) = \hat{I} \cdot \cos(-30°) = i_V(t = 0\,s)$$

Apparently, the currents at the time instant $t = 0$T have a similar relationship as the relative spatial positions between the stator segments and PMs. For instance, the current i_U at the time instant $t = 0$T equals i_V at the time instant $t = 1/3$T.

Another magnetic periodicity can be found by observing the FEM results of PMAFM-ISA at the rotor position 5°, which corresponds to the time instant $t = 1/6\mathrm{T}$, as illustrated in Fig. 4.13.

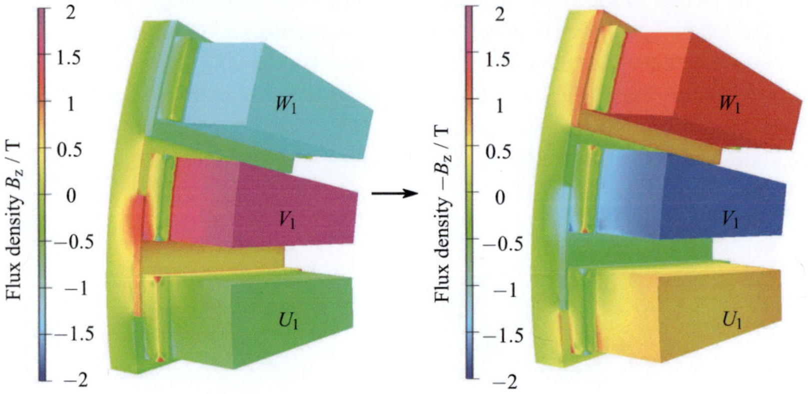

Figure 4.13. Flux density B_z of a PMAFM-ISA at the rotor position 5°

For the rotor position, the PM_1 is aligned with the slot between the stator segments U_1 and V_1, while the PM_2 with opposite magnetization direction is aligned with the stator segment W_1. In comparison, the PM_1 is aligned with the stator segment U_1 and PM_2 is aligned with the slot between the stator segments V_1 and W_1 for the initial rotor position 0°. It implies that the relative spatial positions between the stator segments and PMs for the rotor positions 0° and 5° are identical, except the PM polarity is opposite.

Besides, the current flowing in the coils at the time instants 0T and 1/6T have similar relationship as the relative spatial positions.

$$
\begin{aligned}
i_\mathrm{U}(t = 1/6\mathrm{T}) &= \hat{I} \cdot \cos(150°) = -i_\mathrm{V}(t = 0\,\mathrm{s}) \\
i_\mathrm{V}(t = 1/6\mathrm{T}) &= \hat{I} \cdot \cos(30°) = -i_\mathrm{W}(t = 0\,\mathrm{s}) \\
i_\mathrm{W}(t = 1/6\mathrm{T}) &= \hat{I} \cdot \cos(-90°) = -i_\mathrm{U}(t = 0\,\mathrm{s})
\end{aligned}
$$

For instance, the current i_U at the time instant 0T equals the reverse of i_V at the time instant $t = 1/6\mathrm{T}$.

As a result, the magnetic fields of PMAFM-ISA at rotor positions 0° and 5° have the same amplitude but inverse direction, as illustrated in Fig. 4.13. In summary, it can be noted that the FEA results for the rotor positions $0°(0T)$, $5°(1/6T)$, $10°(1/3T)$, $15°(1/2T)$, $20°(2/3T)$ and $25°(5/6T)$ are either identical or have the same magnitude but inverse direction. The same relationship can be observed for PMAFM-ISA at the other rotor positions as well.

Based on the above analysis, it can be concluded that the electromagnetic field repeats itself after $1/6$ electrical period. Thus, it is only necessary to perform the FEA for $1/6$ electrical period to fully characterize a PMAFM-ISA. The iron losses can be calculated based on the reconstructed flux density waveform in each SMC element for an entire electrical period from the limited available FEA results.

4.3.2 FDW reconstruction for SMC elements in stator

The reconstruction of the FDW for SMC elements in stator is investigated with the above PMAFM-ISA with the DLCW 12 × 2A1. For this purpose, three relevant SMC elements A, B and C of FEA Model in the stator segments U_1, V_1 and W_1 are selected, as illustrated in Fig. 4.14.

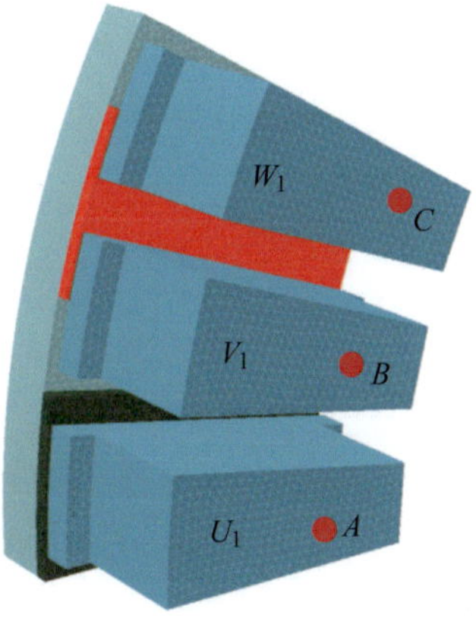

Figure 4.14. Elements in a PMAFM-ISA with DLCW 12 × 2A1

The three SMC elements have the same radial and axial coordinates, while the angle difference among them is 10°.

$$Z_A = Z_B = Z_C$$
$$R_A = R_B = R_C$$
$$\theta_A + 10^\circ = \theta_B = \theta_C - 10^\circ$$

Based on the previously investigated electromagnetic periodicity, it is clear that the FDW of the element *A* for an entire electrical period can be obtained by rearranging the FDW of the elements *A*, *B* and *C* for $1/6$ electrical period according to the following sequence:

$$[\textbf{+U} \quad \textbf{-V} \quad \textbf{+W} \quad \textbf{-U} \quad \textbf{+V} \quad \textbf{-W}]$$

For instance, the reconstructed B_z-FDW of element *A* is validated by comparing with the FDW directly calculated with FEA for an entire electrical period, as shown in Fig. 4.15. It can be seen that the reconstructed FDW matches very well with the directly calculated FDW in element *A*.

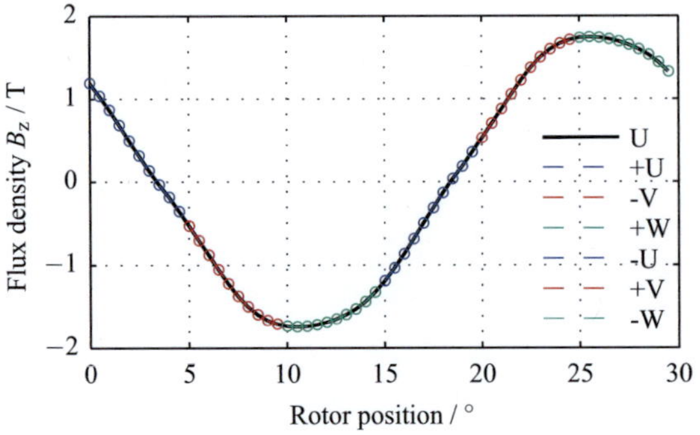

Figure 4.15. Reconstructed B_z-FDW in SMC element *A* in the stator of PMAFM-ISA with DLCW 12 × 2A1

The concrete rearranging sequence is depending on the winding variant of DLCW. Considering the electromagnetic symmetry in PMAFM-ISA with the winding variant *B*, it is only necessary to analyse the model including half of the elementary winding. Subsequently, the reconstruction of the FDW for a PMAFM-ISA with the DLCW 3 × 2B2 is investigated, whose pole pairs number is 15. In similar, three relevant SMC elements A, B and C in the corresponding stator segments U_1, V_1 and W_1 are selected, as illustrated in Fig. 4.16.

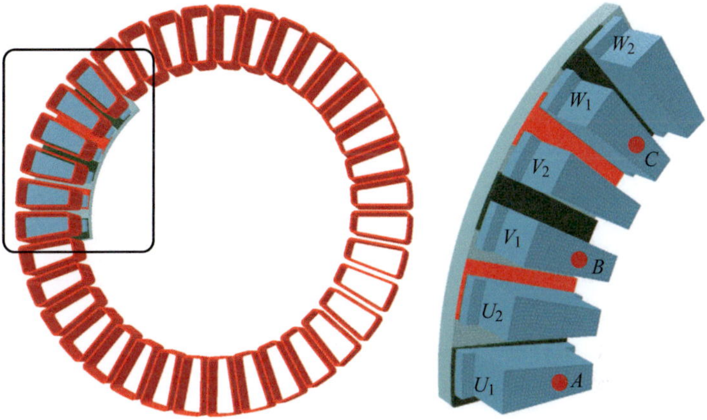

Figure 4.16. FEM Model of a PMAFM-ISA with DLCW $3 \times 2B2$

Through the same analysis, it is noted that the magnetic fields of the PMAFM-ISA for the rotor positions $0°(t = 0T)$, $4°1/6T$ and $8°1/3T$ are identical. In comparison, the magnetic fields for the positions $12°(t = 1/2T)$, $16°(t = 2/3T)$ and $20°(t = 5/6T)$ have the same magnitude but inverse direction. Thus, the rearranging sequence to reconstruct the FDW of SMC element A for an entire electrical period based on the FDW of A, B and C for $1/6$ electrical period is:

$$[\mathbf{+U} \quad \mathbf{+V} \quad \mathbf{+W} \quad \mathbf{-U} \quad \mathbf{-V} \quad \mathbf{-W}]$$

According to the above sequence, the FDW of B_z in SMC element A is reconstructed, which matches well with that directly calculated with 3D FEA, as shown in Fig. 4.17.

In the FEA, the rotational direction of the rotor during the FEA remains unchanged. Therefore, the current in the phases V and W should be exchanged if the rotational direction of working harmonic is different from that of the rotor, so that the fundamental harmonic of the electromotive force and the rotor are kept the same. Both current specifications of PMAFM-ISA are specified in Table 4.1.

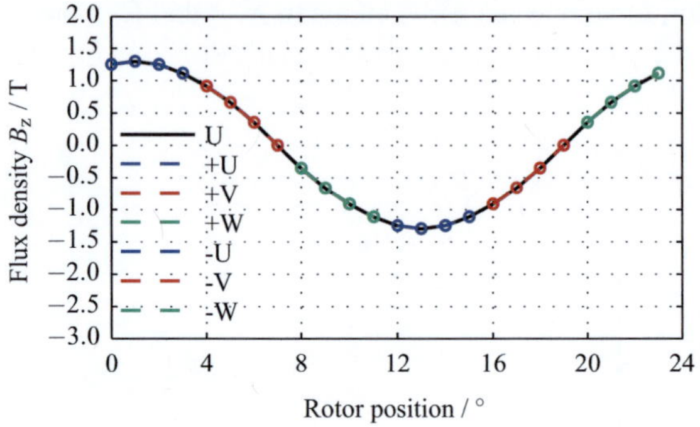

Figure 4.17. Reconstructed B_z-FDW of SMC element A in the stator of PMAFM-ISA with DLCW 3 × 2B2

Current Case	i_U	i_V	i_W
1	$\hat{I} \cdot \cos(wt + 90°)$	$\hat{I} \cdot \cos(wt - 30°)$	$\hat{I} \cdot \cos(wt - 150°)$
2	$\hat{I} \cdot \cos(wt + 90°)$	$\hat{I} \cdot \cos(wt - 150°)$	$\hat{I} \cdot \cos(wt - 30°)$

Table 4.1. Current specifications in PMAFM-ISA in the work

The current specification of both above researched PMAFM-ISAs corresponds to the current case 1 in Table 4.1. For the current case 2, the rearranging sequence should be correspondingly revised. In summary, the general rearranging sequences to reconstruct the FDW for SMC element in the stator are listed in Table 4.2.

Current Case	**Winding Variant**	**Rearranging Sequence**
1	A	$+U - V + W - U + V - W$
	B	$+U + V + W - U - V - W$
2	A	$+U - W + V - U + W - V$
	B	$+U - W - V - U + W + V$

Table 4.2. Rearranging sequence to reconstruct the FDW of SMC element in Stator

At last, it should be noted that for the PMAFM-ISA with winding number N greater than 1 the magnetic fields in the segments belonging to one coil group of a phase are different and cannot be derived from each other. For instance, the magnetic fields in stator segments U_1 and U_2 in Fig. 4.16 are different. Therefore, the FDW in SMC element or rather the iron losses of the segments are not the same and should be calculated independently [S3].

4.3.3 FDW reconstruction for SMC elements in rotor

In order to investigate the reconstruction of the FDW of SMC elements in the rotor, the same PMAFM-ISA with DLCW 12 × 2A1 and 3 × 2B2 are researched. Furthermore, the complete rotor yoke in the FEM model is evenly divided into as many parts as the number of the PMs in the FEM model, as illustrated in Fig. 4.18.

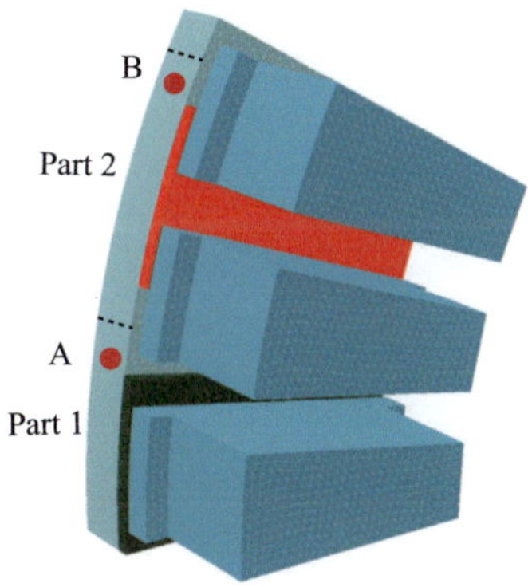

(a) DLCW 12 × 2A1

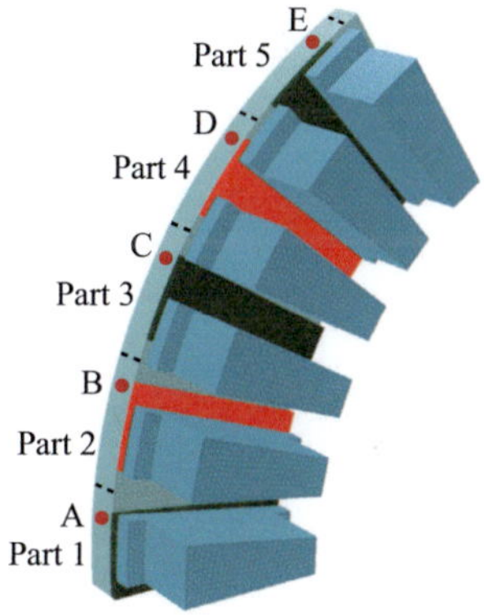

(b) DLCW 3 × 2B2

Figure 4.18. Segmentation of the rotor yoke of PMAFM-ISA with different DLCW

In similar, the relevant SMC elements in the divided parts of the rotor yoke of both PMAFM-ISAs are selected. They have the same radial and axial coordinates, while the angle differences among them are 15° and 12° respectively.

$$\theta_A + 15^\circ = \theta_B$$

$$\theta_A + 24^\circ = \theta_B + 12^\circ = \theta_C = \theta_D - 12^\circ = \theta_E - 24^\circ$$

By observing the magnetic fields of PMAFM-ISA with DLCW $12 \times 2A1$ illustrated in Fig. 4.12 and Fig. 4.13, it is clear that the magnetic fields of the rotor yoke repeats itself after $1/6$ electrical period and the period of the FDW of the SMC element A in part 1 is equal to $10°$.

The FDW of element A can be reconstructed by rearranging the FDW of elements A and B corresponding to $1/6$ electrical period according to the sequence

$$\left[\textbf{+Part 1} \quad \textbf{-Part 2} \right]$$

For instance, the reconstructed B_z-FDW of SMC element A in PMAFM-ISA with DLCW $12 \times 2A1$ is compared with that directly determined with FEA, as illustrated in Fig. 4.19. It can be seen that there is very small deviation.

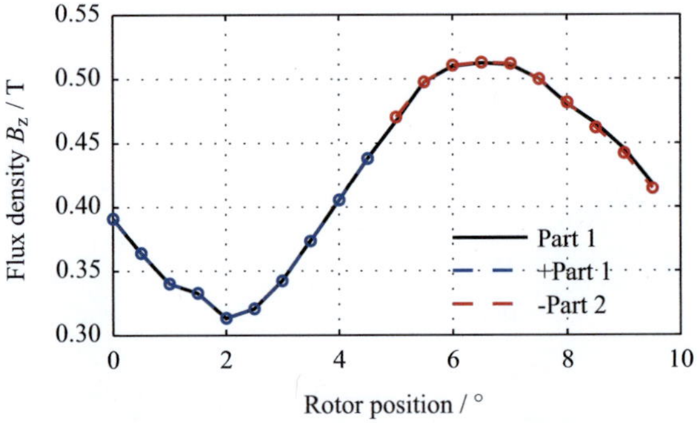

Figure 4.19. Reconstructed B_z-FDW of SMC element A in the rotor of PMAFM-ISA with DLCW $12 \times 2A1$

For the PMAFM-ISA with DLCW $3 \times 2B2$ in Fig. 4.18, the FDW in element A can be reconstructed by rearranging the FDW of elements A, B, C, D and E corresponding to $1/6$ electrical period. In order to determine the rearranging sequence for reconstruction of FDW, the 3D FEM model is linearised at the mean radius for a better view, as illustrated in Fig. 4.20. Hereby, the rotor yoke is divided into five segments with surface mounted five PMs.

First of all, a vector P_1 representing the positions of the stator segments U_1, V_1 and W_1 should be defined, whose length is equal to the number of PMs on the rotor yoke.

In addition, the sequence should be determined considering winding variant and current specification in Table 4.1, which is the same as the sequence that current in the coils around the segments is zero when looking backwards.

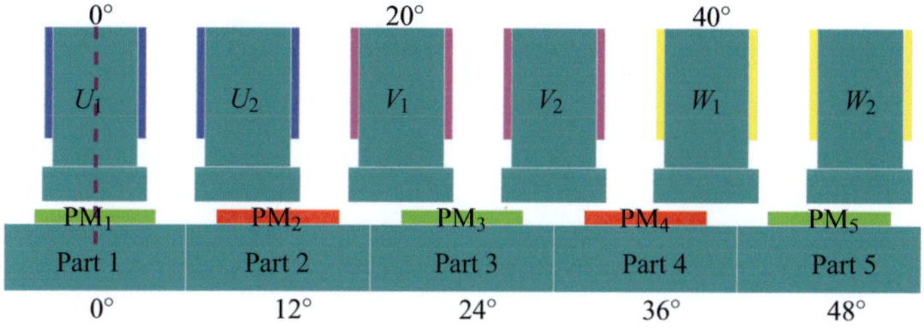

Figure 4.20. Linearised model of PMAFM-ISA with DLCW 3 × 2B2

For the researched PMAFM-ISA with DLCW 3 × 2B2, the sequence that current in the coils around the segments U_1, V_1 and W_1 equalling zero is

$$\begin{bmatrix} \cdots U_1 & W_1 & \overleftarrow{V_1 \quad U_1 \quad W_1 \quad V_1 \quad U_1} & W_1 & V_1 \cdots \end{bmatrix}$$

As a result, the corresponding vector P_1 of the stator segments is

$$P_1 = [0^\circ \quad 20^\circ \quad 40^\circ \quad 0^\circ \quad 20^\circ]$$

Subsequently, the other vector P_2 representing the positions of the five PMs in the FEM model is defined as

$$P_2 = [0^\circ \quad 12^\circ \quad 24^\circ \quad 36^\circ \quad 48^\circ]$$

As the number of pole pairs of PMAFM-ISA equals 15, the FEA over a rotation of 4° corresponding to $1/6$ electrical period is carried out. It is assumed that the PM_1 on **part 1** is aligned with the first stator segment U_1 at initial rotor position. After rotation corresponding to $1/6$ electrical period, there is a relative angular deviation of 4° between the PM_1 and U_1, which equals the deviation between the PM_3 on **part 3** and V_1 (the second element in the vector P_1) at the initial rotor position ($t = 0\,\text{s}$). Subsequently, after the rotation of $2/6$ electrical period, the angular difference between PM_1 and U_1 becomes 8°, which equals the deviation between the PM_5 on **part 5** and W_1 (the third element in the vector P_1) at initial time instant ($t = 0\,\text{s}$).

In similar, it can be noted that the angular difference between PM_1 and U_1 is equal to 12° after rotation of 3/6 electrical period, which equals the deviation between PM_2 on **part 2** and segment U_1 (the fourth element in vector P_1) at initial time instant ($t = 0\,\mathrm{s}$). At last, the difference between PM_1 and U_1 after rotation of 4/6 electrical period is 16°, which equals the deviation between PM_4 on **part 4** and V_1 (the fifth element of vector P_1) at initial rotor position.

Furthermore, it should be noted that PM_2 and PM_4 have inverse magnetization direction compared with PM_1, PM_3 and PM_5. Based on the above consideration, it is apparent that the rearranging sequence to reconstruct the FDW of SMC element A based on the FDW in elements A, B, C, D and E for 1/6 electrical period is:

[**+Part 1 +Part 3 +Part 5 -Part 2 -Part 4**]

For instance, the B_t-FDW in SMC element A in part 1 is reconstructed based on the FEA over 1/6 electrical period according to the previously determined rearranging sequence for $I = 0\,\mathrm{A}$ and $I \neq 0\,\mathrm{A}$ respectively. For a better comparison, the FDW directly calculated from FEA for an entire electrical period is also illustrated in Fig. 4.21, which matches very well with the reconstructed FDW.

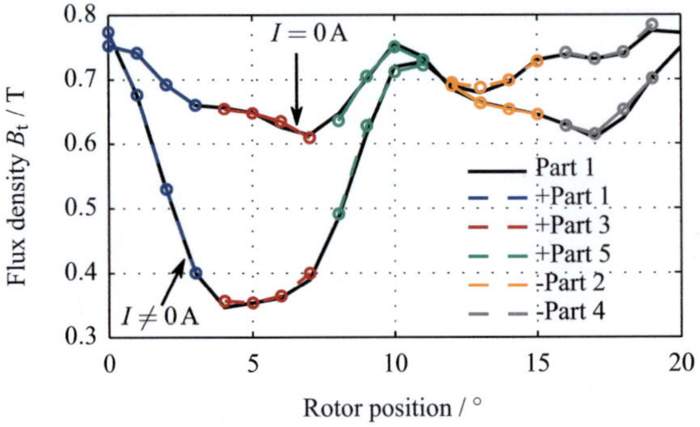

Figure 4.21. Reconstructed B_t-FDW of PMAFM-ISA with DLCW 3 × 2B2

In general, the rearranging sequence can be determined according to the flow chart illustrated in Fig. 4.22. First of all, the vector P_1 representing the positions of stator segments and the vector P_2 representing the positions of PMs in the FEM model should be determined. Subsequently, for each element of P_1, the suitable element in P_2 should be determined. For the i-th element of the vector

P_1, there should be an angular difference equalling $i \cdot 60°/p$ between them. In addition, if the selected PM has an opposite magnetization direction, the flux density sign of the SMC elements should be inverted as well.

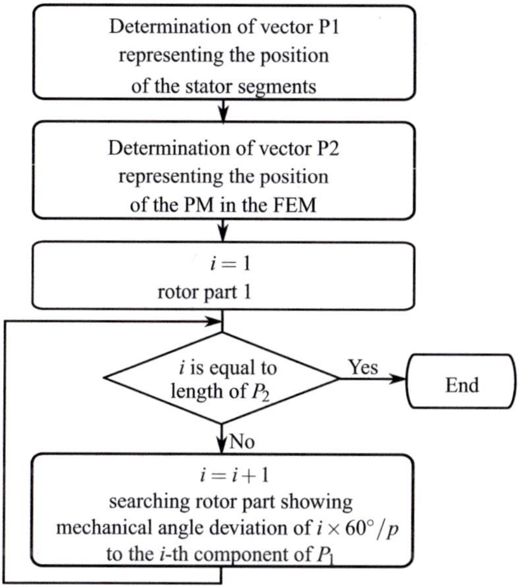

Figure 4.22. Flow chart determining rearranging sequence of rotor segments

By observing the reconstructed FDW in Fig. 4.21, it can be noted that the period of the SMC element of rotor yoke is depending on the electrical loading for the PMAFM-ISA with DLCW 3 × 2B2. The periods are 10° and 20° , when the PMAFM-ISA works without and with electrical loading respectively.

In general, the frequency of FDW of SMC elements in rotor yoke f_{rotor} is different from that of SMC elements in stator f_{stator}. In other words, for the iron loss model described in section 3.3, the frequency should be determined considering the position of the SMC elements and the electrical loading of PMAFM-ISA. For the SMC element of stator segments, the frequency can be calculated from the rotational speed n in min^{-1} and the number of pole pairs p using (4.10).

$$f_{\text{stator}} = \frac{n}{60} \cdot p \tag{4.10}$$

In comparison, the period of FDW of SMC element in rotor yoke for PMAFM-ISA at no-load can be calculated using (4.11), where N_s is the number of the stator segments.

$$\theta_{\text{rotor}} = \frac{360^\circ}{N_s} \tag{4.11}$$

Therefore, the corresponding frequency of the SMC element in the rotor yoke in this case is:

$$f_{\text{rotor},i=0} = f_{\text{stator}} \cdot \frac{N_s}{p} \tag{4.12}$$

If PMAFM-ISA works with electrical loading, the period of FDW of SMC element in rotor yoke can be calculated with (4.13), where N is winding number equalling the adjacent coils belonging to a coil group of a phase.

$$\theta_{\text{Rotor}} = \frac{360^\circ}{N_s} \cdot N \tag{4.13}$$

For this case, the frequency of SMC element in rotor yoke is equal to

$$f_{\text{rotor},i\neq 0} = f_{\text{stator}} \cdot \frac{N_s}{p \cdot N} \tag{4.14}$$

In summary, a method to reconstruct the complete FDW of SMC elements from limited FEA results of $1/6$ electrical period is described in this section. For the SMC element in stator and rotor yoke, the determination of the rearranging sequence is explained respectively, which can be well validated by comprising the reconstructed FDW and the FDW directly determined with FEA for an entire electrical period. Furthermore, the frequency of the iron loss model described in section 3.3 should be determined depending on the position of the SMC element and the electrical loading of PMAFM-ISA.

5

Electromagnetic Design of PMAFM-ISA

With the help of the developed simulation platform based on 3D FEM, many promising topologies of electrical machine can be investigated. The permanent magnet excited axial flux machine with internal segmented armature (PMAFM-ISA) illustrated in Fig. 5.1 is an attractive and promising topology for the application of SMC. It posses high torque and power density due to the application of PM with high energy density and the concentrated winding with short end-windings. In addition, the absence of the stator yoke leads to significant reduction of the stator core weight and iron losses [S4].

The chapter is organized as follows. At first, the basic theory of PMAFM-ISA and two different methods based on the analytical calculation to determine the torque are presented. Subsequently, they are compared with the 2D and 3D FEM. On this basis, the basic electromagnetic design of PMAFM-ISA considering the design specifications has been carried out, which is then optimized with the previously developed simulation platform. At last, the final electromagnetic design of PMAFM-ISA is presented and the demagnetization of PM is carefully examined as well.

The PMAFM-ISA consists of two external rotors with surface mounted PMs and an internal stator comprising several independent segments with concentrated winding. Considering the electromagnetic symmetry and periodicity described in sections 4.2 and 4.3, it is only necessary to analyse the model comprising one elementary winding for the winding variant A or half of the elementary winding for the winding variant B. For instance, it is only necessary to analyse 1/24 of a complete PMAFM-ISA with the DLCW $12 \times 2A1$, whose meshed model and the FEM results at no-load are illustrated in Fig. 5.1.

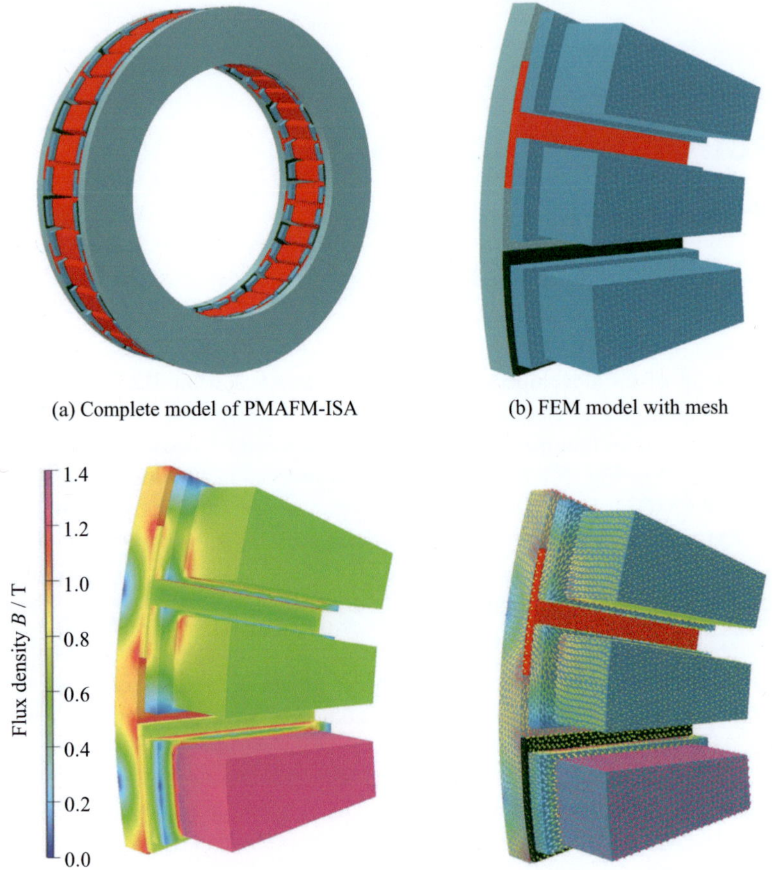

(a) Complete model of PMAFM-ISA (b) FEM model with mesh

(c) *B*-Field (d) Magnetic flux

Figure 5.1. Structure and FEM results of a PMAFM-ISA

It is apparent that the magnetic flux flows three-dimensionally in the rotor yoke and stator segments. For this reason, the SMC are used to manufacture the complex components. By this means, the problems associated with punching and winding resulted from the usage of punched spirally wound laminations can be well avoided. On the other side, the manufacturing process of complex stator segments with end extrusion can be significantly simplified, so that the utilization of both space and the flux generated by PM are improved.

5.1 Analysis based on magnetic network and FEM

There are many different methods to analyse PMAFM-ISA. In this section, three different methods are presented and compared, which are the high efficient analytical calculation, 2D FEM and 3D FEM respectively. For a better comparison, the model in Fig. 5.1 has been analysed with the three methods, whose parameters are listed in Table 5.1.

Symbol	**Description**	**Value**	**Unit**
D_{o}	active outer diameter	301	mm
D_{i}	active inner diameter	209	mm
L_{motor}	active total axial length	75	mm
N_{s}	number of stator segments	36	-
p	number of pole pairs	12	-
L_{g}	axial length of air gap	1	mm
L_{shoe}	axial length of stator shoe	5	mm
L_{bar}	axial length of stator bar	39.6	mm
L_{yoke}	axial length of rotor yoke	9.5	mm
L_{pm}	axial length of PM	2.2	mm
m_{phase}	number of phases	3	-
a_{path}	number of parallel branches of each phase	6	-
B_{coil}	width of the coil around a segment	3.82	mm
w_{coil}	number of turns per coil around a segment	30	-
α_{pm}	ratio of PM arc to pole pitch	2/3	-

Table 5.1. Main parameters of the PMAFM-ISA in Fig. 5.1

5.1.1 Analytical analysis

Compared with the time-consuming FEM, the analytical calculation based on the magnetic network is a fast and efficient method, which can be developed by utilizing the analogy to electrical circuit. The basic magnetic network is based on the linearised cross section of the PMAFM-ISA at geometrical mean radius, which is illustrated in Fig. 5.2.

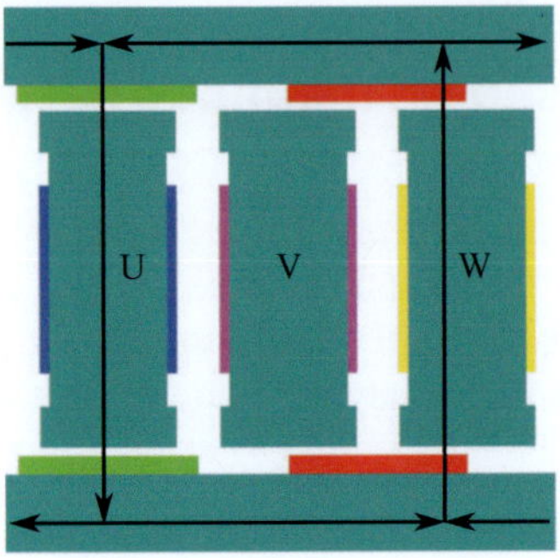

Figure 5.2. Simplified magnetic network of PMAFM-ISA

Ignoring the magnetic reluctance of SMC components, leakage flux and slot effect, the magnetic flux density in the air gap at no-load has a rectangular waveform covering 120 electrical degrees or $2/3$ pole pitch, whose amplitude B_{m} can be calculated using (5.1)

$$B_{\mathrm{m}} = \frac{B_{\mathrm{rpm}} \cdot L_{\mathrm{pm}}}{\mu_{\mathrm{rpm}} \cdot L_{\mathrm{g}} + L_{\mathrm{pm}}} \tag{5.1}$$

where L_{g}, B_{rpm}, μ_{rpm} and L_{pm} are axial length of the air gap, remanent flux density, relative permeability and axial length of the PM respectively.

In some literatures, the Carter coefficient k_{c} is introduced to take the influence of the slot opening into account, which leads to decrease of the flux density in the air gap under each slot opening due to the increasing reluctance. The decreased flux density amplitude B_{m} can be calculated using (5.2).

$$B_{\mathrm{m}} = \frac{B_{\mathrm{rpm}} \cdot L_{\mathrm{pm}}}{\mu_{\mathrm{rpm}} \cdot k_{\mathrm{c}} \cdot L_{\mathrm{g}} + L_{\mathrm{pm}}} \tag{5.2}$$

There are many different methods to calculate the Carter coefficient k_c. One of them is defined in (5.3), where B_s is width of slot opening and τ_{slot} is slot pitch.

$$k_c = \frac{\tau_{slot}}{\tau_{slot} - L_g \cdot \frac{4}{\pi} \left[\frac{B_s}{2L_g} \arctan\left(\frac{B_s}{2L_g}\right) - \ln\sqrt{1+\left(\frac{B_s}{2L_g}\right)^2} \right]} \tag{5.3}$$

Subsequently, the peak value of the fundamental harmonic $\hat{B}_1$ can be calculated through the Fourier analysis using (5.4).

$$\hat{B}_1 = \frac{4\sqrt{3}}{2\pi} \cdot B_m \tag{5.4}$$

Another method to calculate the no-load magnetic field distribution in the air gap is based on the solution of the scalar magnetic potential in both air and PM governed by Laplace's equations [79, 91, 107, 133, 138]. Hereby, the slot effect and the reluctance of SMC are also ignored. According to this method, the flux density component generated by PM along the axial direction B_{gpm} can be calculated using (5.5), where τ_p is pole pitch.

$$\begin{aligned}
B_{gpm} &= \sum_{n=1,3\cdots}^{\infty} \frac{B_{pm1}}{B_{pm2}} \\
B_{pm1} &= \frac{8\,B_{rpm}}{n\,\pi} \sin\left(\frac{\alpha_{pm}\,n\,\pi}{2}\right) \exp\left(\frac{-n\,\pi\,L_g}{\tau_p}\right) \cos\left(\frac{n\,\pi\,x}{\tau_p}\right) \\
B_{pm2} &= \frac{\mu_{rpm}\left(-\exp\left(\frac{-2\,n\,\pi\,L_g}{\tau_p}\right)+1\right)\left(\exp\left(\frac{-2\,n\,\pi\,L_g}{\tau_p}\right)+1\right)}{\mu_0\left(\exp\left(\frac{-2\,n\,\pi\,L_g}{\tau_p}\right)-1\right)} \\
&\quad + \exp\left(\frac{-2\,n\,\pi\,L_g}{\tau_p}\right)+1
\end{aligned} \tag{5.5}$$

Considering the influence of slot opening, the B_{gpm} should be multiplied with a coefficient λ_s. As illustrated in Fig. 5.3, the coefficient λ_s can be calculated using (5.6), where r_s is the equivalent radius of the magnetic flux. It is equal to 1 under the stator tooth and smaller than 1 under the slot opening.

$$\lambda_s = \frac{L_g + \dfrac{L_{pm}}{\mu_{rpm}}}{L_g + \dfrac{L_{pm}}{\mu_{rpm}} + \dfrac{2\,\pi\,r_s}{4}} \tag{5.6}$$

Figure 5.3. Influence of slot opening

A comparison of the flux density in the air gap of the PMAFM-ISA at no-load is illustrated in Fig. 5.4, which are calculated from the simple method considering the Carter coefficient, the analytical method based on solving of Laplace's equations with consideration the influence of slot opening and the 3D FEM respectively. It can be noted that the analytically calculated flux density is slightly higher than FEM, as the SMC reluctance is ignored.

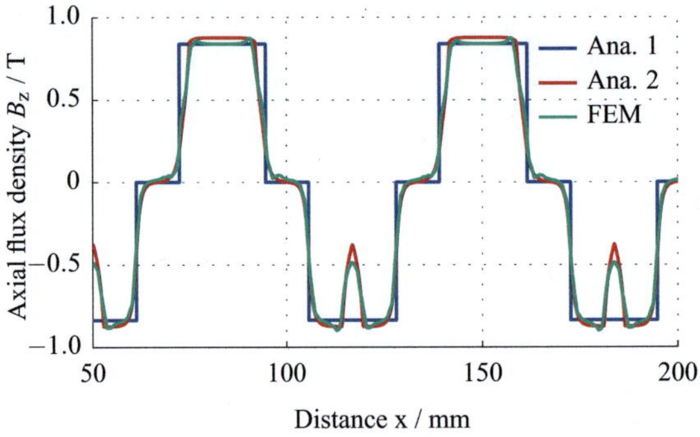

Figure 5.4. Comparison of the calculated flux density of PMAFM-ISA at no-load

The magnetic flux generated by a PM Φ_{pm} can be calculated as the product of B_m and its area A_{pm} using (5.7), where R_o and R_i are the outer and inner radius of PMAFM-ISA.

$$\Phi_{pm} = B_m \cdot A_{pm} = \alpha_{pm} \cdot B_m \cdot \pi \cdot \frac{R_o^2 - R_i^2}{2 \cdot p} \tag{5.7}$$

As the magnetic flux Φ_{pm} is divided into two parts with opposite flowing direction in the rotor yoke, the magnetic flux density in the rotor yoke B_{yoke} can be calculated using 5.8, where L_{yoke} is the axial length of rotor yoke.

$$B_{yoke} = \frac{\Phi_{pm}}{2 \cdot (R_o - R_i) \cdot L_{yoke}} = \alpha_{pm} \cdot \frac{B_m}{L_{yoke}} \cdot \frac{\pi}{4 \cdot p} (R_o + R_i) \tag{5.8}$$

By observing the concentrated winding around a stator segment illustrated in Fig. 5.5, it can be noted the area of the coil around the stator bar S_{coil} can be approximated as the sum of 4 rectangles. Its area can be calculated using (5.9), where B_{coil} and N_s are the width of the coil around a stator segment and the total number of stator segments.

$$S_{coil} = B_{coil} \cdot \left(2(R_o - R_i) + (\frac{\pi \cdot D_o}{N_s} - 2 \cdot B_{coil}) + (\frac{\pi \cdot D_i}{N_s} - 2 \cdot B_{coil}) \right) \tag{5.9}$$

Figure 5.5. Cross section of the coil around a stator segment

On this basis, the length and the area of stator bar L_{bar}, S_{bar} can be calculated using the following (5.11):

$$L_{\mathrm{bar}} = L_{\mathrm{motor}} - 2 \cdot (L_{\mathrm{yoke}} + L_{\mathrm{pm}} + L_{\mathrm{g}} + L_{\mathrm{shoe}}) \quad (5.10)$$

$$S_{\mathrm{bar}} = \frac{\pi \cdot (R_{\mathrm{o}}^2 - R_{\mathrm{i}}^2)}{N_{\mathrm{s}}} - S_{\mathrm{coil}} \quad (5.11)$$

The flux density in stator bar B_{bar} can be calculated from the magnetic flux through the stator bar and its area using (5.12).

$$B_{\mathrm{bar}} = \frac{\Phi_{\mathrm{pm}}}{S_{\mathrm{bar}}} = \alpha_{\mathrm{pm}} \cdot \frac{B_{\mathrm{m}}}{S_{\mathrm{bar}}} \cdot \pi \frac{R_{\mathrm{o}}^2 - R_{\mathrm{i}}^2}{2 \cdot p} \quad (5.12)$$

Based on the above analysis, the torque can be calculated from the magnetic loading B_{load}, electrical loading Q_{load} and the area of air gap with (5.13).

$$T = \pi \cdot B_{\mathrm{load}} \cdot Q_{\mathrm{load}} \cdot \frac{D_{\mathrm{o}} + D_{\mathrm{i}}}{2} \cdot D_{\mathrm{i}} \cdot \frac{D_{\mathrm{o}} - D_{\mathrm{i}}}{2} \quad (5.13)$$

The magnetic loading B_{load} is equal to the rms-value of the fundamental harmonic of flux density wave form:

$$B_{\mathrm{load}} = \frac{\hat{B}_1}{\sqrt{2}} \quad (5.14)$$

On the other hand, the electrical loading Q_{load} can be calculated based on the current density or the phase current using (5.15) or (5.16), where k_{w1} is the winding factor of the fundamental harmonic, m_{phase} is the number of phases, k_{coil} is the filling factor of coil and n_{series} is the number of coils of one phase connected in series.

$$Q_{\mathrm{load}} = k_{\mathrm{w1}} \cdot \frac{m_{\mathrm{phase}} \cdot I_{\mathrm{phase}} \cdot w_{\mathrm{coil}} \cdot n_{\mathrm{series}}}{\pi \cdot D_{\mathrm{i}}} \quad (5.15)$$

$$Q_{\mathrm{load}} = k_{\mathrm{w1}} \cdot \frac{m_{\mathrm{phase}} \cdot J_{\mathrm{rms}} \cdot A_{\mathrm{coil}} \cdot k_{\mathrm{coil}} \cdot n_{\mathrm{series}}}{\pi \cdot D_{\mathrm{i}}} \quad (5.16)$$

The A_{coil} represents the available cross sectional area of coil around a stator segment, which can be evaluated as

$$A_{\mathrm{coil}} = L_{\mathrm{bar}} \cdot B_{\mathrm{coil}} \tag{5.17}$$

However, the above method leads to great deviation at high electrical loading due to the simplification in the magnetic network. In order to improve the accuracy, the above magnetic network is further improved in terms of following aspects: [S5]:

- Consideration of the magnetic reluctance of SMC and its non-linearity, which means the reluctance is not constant but depending on the saturation level.
- Consideration of the true magnetic flux path including the leakage flux.
- Segmenting the stator segment into many small components to approximate the magnetic field more accurately.

The improved equivalent magnetic network has been developed with the toolbox "Simscape" in Matlab/Simulink for the PMAFM-ISA with DLCW 12 × 2A1 and 3 × 2B2. For the former, the improved equivalent magnetic network corresponds to the initial rotor position where the phase U is aligned with the permanent magnet, as illustrated in Fig. 5.6.

Hereby, the PM is modelled as an ideal magnetomotive force (MMF) with inner reluctance, while the coil around the stator segment is modelled as a controlled MMF source, whose value is depending on the current in the coil and the number of turns per coil. In addition, the SMC component is modelled as variable reluctance, which is depending on the equivalent length, the cross sectional area of the SMC component and the relative permeability corresponding to the saturation level.

With the help of the magnetic flux sensor, the magnetic flux through the three stator segments ϕ_{u}, ϕ_{v} and ϕ_{w} can be determined. Considering that the axis of phase U is aligned with the d-axis of the rotor for the current position and there is only current component i_{q}, the magnetic flux Ψ_{d} and the torque can be calculated as follows:

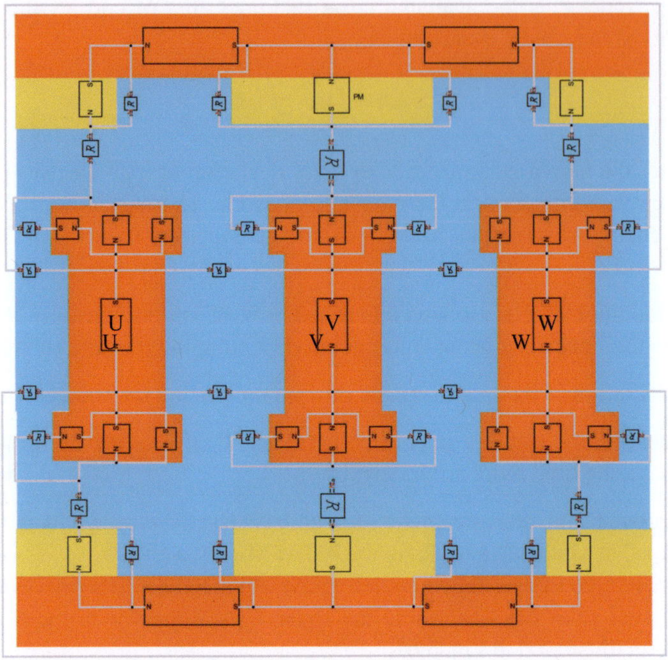

Figure 5.6. Improved equivalent magnetic network of a PMAFM-ISA

$$\Psi_{\mathrm{d}} = w_{\mathrm{coil}} \cdot n_{\mathrm{series}} \cdot \frac{2}{3} \cdot \left(\phi_{\mathrm{u}} - \frac{1}{2}\phi_{\mathrm{v}} - \frac{1}{2}\phi_{\mathrm{w}} \right) \tag{5.18}$$

$$T = \frac{3}{2} \cdot p \cdot \left(\Psi_{\mathrm{d}} \cdot I_{\mathrm{q}} \right) \tag{5.19}$$

5.1.2 Finite element analysis

In addition to the analytical calculation, the PMAFM-ISA can be analysed with both 2D and 3D FEM. The FEM is a numerical method dividing the electrical machine into a large number of small elements. Through solving the governing partial differential equations in each element derived from Maxwell's equations [S6], the electromagnetic field in the electrical machine can be determined.

Generally speaking, the FEM is much more accurate and frequently used to verify the results of analytical methods. Besides the 3D FEM model illustrated in Fig. 5.1, the PMAFM-ISA can be modelled in 2D FEM as well, which is similar to the analytical method based on the linearised cross section at the geometrical mean radius equal to $(R_\mathrm{o} + R_\mathrm{i})/2$, as illustrated in Fig. 5.7.

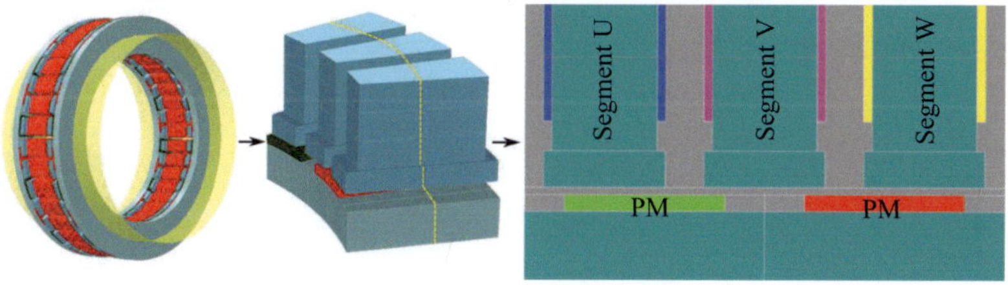

Figure 5.7. 2D FEM model of PMAFM-ISA

For the case of 2D FEM, the rotation of rotor is transformed into the translation of rotor with the velocity V_Trans, which can be calculated from the mechanical rotational speed Ω_Rot using (5.20).

$$V_\mathrm{Trans} = \Omega_\mathrm{Rot} \cdot \frac{R_\mathrm{o} + R_\mathrm{i}}{2} \tag{5.20}$$

A comparison of the B-Filed of a PMAFM-ISA calculated from 2D and 3D FEM respectively for $i_\mathrm{d} = 0\,\mathrm{A}$ and $i_\mathrm{q} = 420\,\mathrm{A}$ are illustrated in Fig. 5.8. It can be noted that some aspects such as the rotor yoke area with low flux density and the tip of stator segment w with high flux density can be well investigated with both methods. However, some aspects such as the area difference between the bar and the shoe of stator segment cannot be correctly taken into account with 2D FEM. As a result, the stator bar has a large cross sectional area in 2D FEM, which leads to lower flux density, lower saturation level and higher torque.

All the above methods are compared in terms of torque, as illustrated in Fig. 5.9. It can be seen that the analytical method based on (5.13) offers the highest torque, even though the carter coefficient is taken into account. The most important reason is that the reluctance of SMC is ignored in the simple magnetic network illustrated in Fig. 5.2. Furthermore, the torque calculated with 2D FEM is also higher than 3D FEM, which is mainly resulted from the larger cross sectional area of stator bar.

(a) 2D FEM

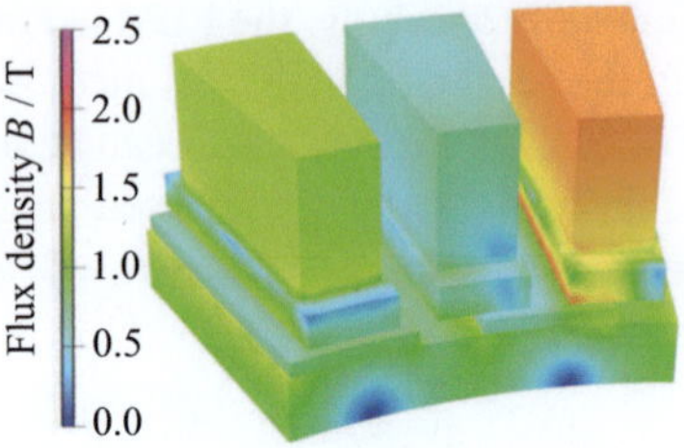

(b) 3D FEM

Figure 5.8. B-Field of a PMAFM-ISA excited with $i_{\mathrm{d}} = 0\,\mathrm{A}$ and $i_{\mathrm{q}} = 420\,\mathrm{A}$

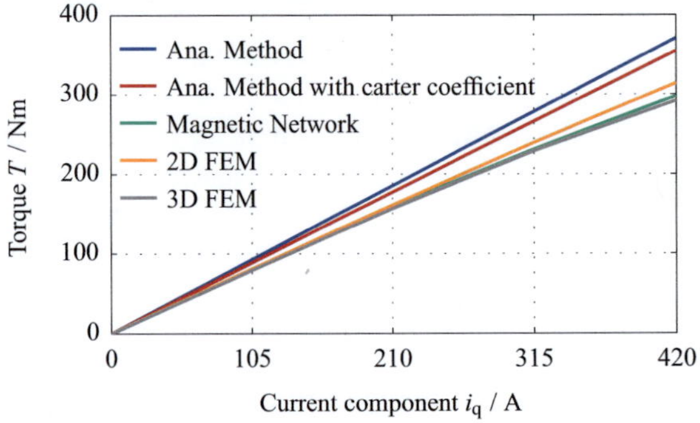

Figure 5.9. Comparison of the torque of a PMAFM-ISA with $i_{\mathrm{d}} = 0\,\mathrm{A}$

In comparison, the deviation between the 3D FEM and the method based on the improved equivalent magnetic network illustrated in Fig. 5.6 is the smallest. However, the disadvantages of the method should also be taken into account. As the equivalent magnetic network depends strongly on flux path, type of concentrated winding, concrete geometric parameters and rotor position, the model is very inflexible. Thus, it must be adjusted and verified with 3D FEM for different PMAFM-ISA. In addition, the available informations based on the improved equivalent magnetic network is very limited and some important characteristics such as iron losses cannot be determined accurately. Due to the inherent 3D flux distribution in AFM, 3D FEM is necessary for the parameter evaluation and the accurate analysis of magnetic field distribution.

Based on the above analysis, it is decided that the electromagnetic design of PMAFM-ISA is carried out in two steps. At first, the basic magnetic design is performed based on the improved equivalent magnetic network illustrated in Fig. 5.6. Subsequently, the basic geometry is further optimized with the 3D FEM as many critical parameters cannot be accurately analyzed in the first step.

5.2 Basic electromagnetic design of PMAFM-ISA

In this section, the geometrical parameters of PMAFM-ISA are determined based on the specified design constraints listed in Table 1.2 and some primitive assumptions.

5.2.1 Number of stator segments and pole-pairs

With the help of the winding editor described in 4.2, it is able to analyse and compare the different layouts of DLCW. Hereby, special attentions have been paid to the winding factor of the fundamental harmonic k_{w1} and the leakage factor σ_d. The two important properties of some promising DLCWs with different elementary windings are listed in Table 5.2.

Elementary winding of DLCW	Winding factor k_{w1}	Leakage factor σ_{d}
2A1	0.866	46%
2B1	0.5	10%
2A2	0.75	95%
2B2	0.933	96.4%
2A3	0.945	117.8%
2B3	0.793	136.6%
2A4	0.808	161.9%
2B4	0.949	128.9%

Table 5.2. Winding factor k_{w1} and leakage factor σ_{d} of some DLCWs

Considering the above listed winding factor k_{w1} and leakage factor σ_{d}, the DLCW with elementary winding 2A1 and 2B2 are the most promising candidates, whose harmonic contents are illustrated in Fig. 5.10.

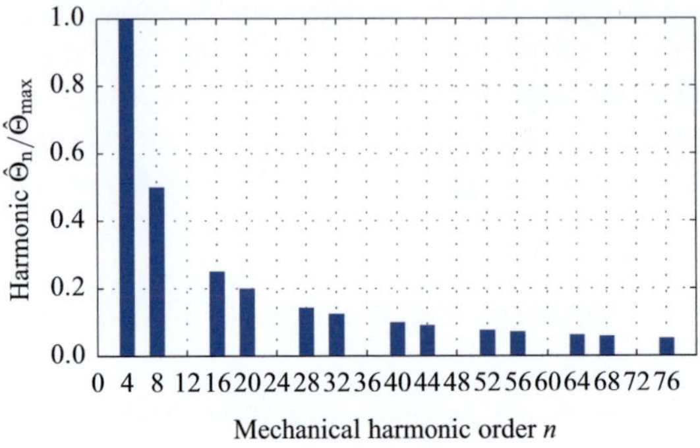

(a) DLCW 4 × 2A1

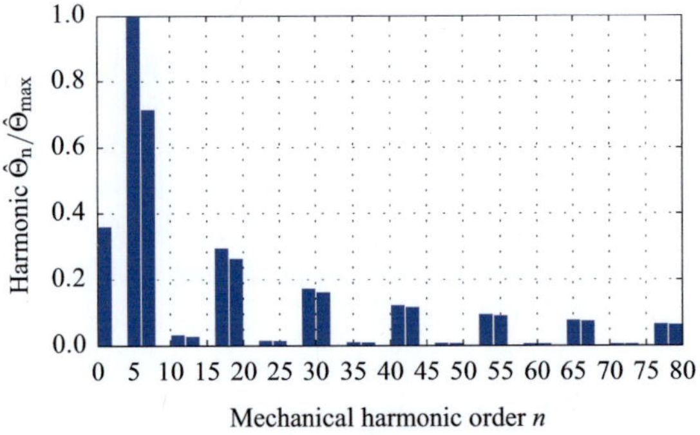

(b) DLCW 1 × 2B2

Figure 5.10. Harmonics contained in two promising DLCWs

In general, the harmonic with the highest amplitude are selected as working harmonic. For instance, the 4-th and 5-th harmonics are chosen to generate torque for PMAFM with 12 stator segments and above DLCWs. It implies that the number of pole pairs p are equal to 4 and 5 respectively. As a result, the harmonic with the lowest order is the working harmonic for PMAFM-ISA with the DLCW 4 × 2A1, while the second harmonic is the working harmonic for PMAFM-ISA with the DLCW 1 × 2B2.

If an electrical machine is fed with inverter, the phase current isn't ideal sinusoidal and contains many harmonics, which can result in additional iron losses in SMC and higher eddy current losses in PM [249, 251, 253–257].

Especially the subharmonics can cause more significant non-synchronous magnetic fields and thus more eddy current losses in PM [247, 251, 258–261]. Furthermore, the PMs are normally glued on the rotor yoke with adhesive, whose maximum mechanical strength is extremely sensitive to the temperature. Based on the above consideration, the DLCW with the elementary winding 2A1 is ultimately selected for the PMAFM-ISA in this work.

Subsequently, the number of the elementary winding U should be determined based on the following consideration:

- First of all, a smaller number of pole pairs leads to larger PM size, which results in a thicker rotor yoke for the specified maximum allowable flux density according to (5.7) and (5.8). In addition, the equivalent length of the magnetic flux in the rotor yoke between two adjacent PMs becomes also longer, which causes larger reluctance. Its influence is considerable because of the low permeability of SMC. It can be concluded that a PMAFM-ISA with lower number of pole pairs can offer lower torque for specified current density.
- Furthermore, a PMAFM-ISA with higher number of pole pairs has higher iron losses for the same rotational speed. Furthermore, the sampling frequency of the inverter increases as well leading to higher losses of the inverter.
- Subsequently, a large number of pole pairs leads to reduction of the end-winding space, which is important when a short axial length is required.
- At last, it should be noted that the available size of the pressed standard SMC cylinder is limited, from which the stator segments are milled. When the number of stator segments is too small, it causes an unacceptable size beyond that of the standard SMC cylinder. In contrast, an unsuitable larger number of stator segments corresponds to a smaller size, which causes more manufacturing and winding difficulties.
- Considering the magnetic symmetry, the 3D FEA of PMAFM-ISA with increasing number of elementary windings needs less calculation time for the same basic element size.

Based on the above consideration, the number of the elementary winding U is determined as 12, which corresponds to 36 stator segments and 12 pole pairs. On this basis, the design process of the PMAFM-ISA is illustrated in Fig. 5.11.

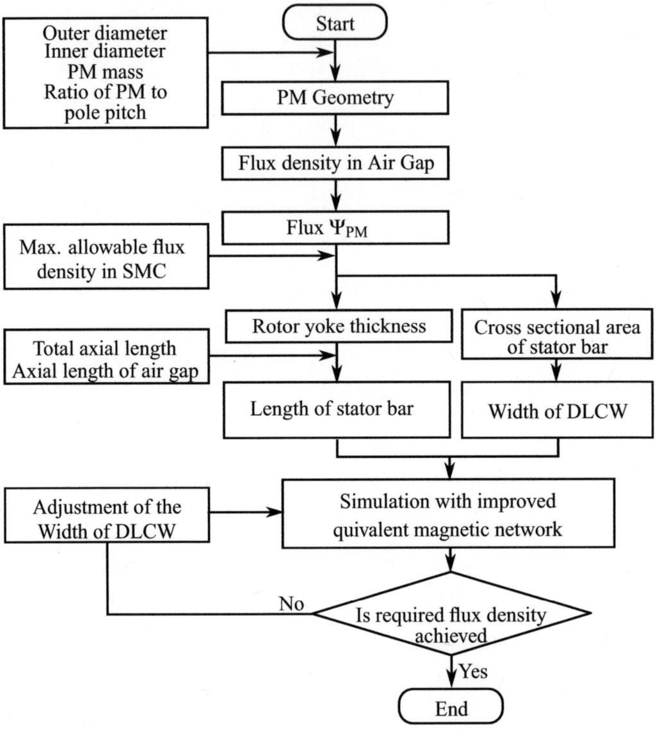

Figure 5.11. Basic design process of PMAFM-ISA

According to the design process, the basic electromagnetic design of PMAFM-ISA has been determined under the following assumptions. First of all, the ratio of PM to pole pitch is equal to $2/3$ and the slot opening factor α_{slot} is 0.7, which is defined as the ratio of the slot opening to the distance of adjacent stator segments. On the other side, the maximum allowable flux density in SMC of PMAFM-ISA at no-load has been set as 1.5 T.

The parameters of the basic electromagnetic design are listed in Table 5.3. Considering relatively low mechanical strength of SMC and its manufacturing difficulty, the minimum thickness of rotor yoke is specified as 9.5 mm.

As the starting point of the electromagnetic design, the SMC material Somaloy 700HR3P is specified for the rotor yoke and stator segments.

Symbol	Description	Value	Unit
D_o	active outer diameter	301	mm
D_i	active inner diameter	209	mm
L_{motor}	active total axial length	75	mm
m_{phase}	number of phases	3	-
N_s	number of stator segments	36	-
p	number of pole pairs	12	-
L_g	axial length of air gap	1	mm
L_{shoe}	axial length of stator shoe	5	mm
L_{bar}	axial length of stator bar	39.6	mm
L_{yoke}	axial length of rotor yoke	9.5	mm
L_{pm}	axial length of PM	2.2	mm
α_{pm}	pole arc to pole pitch ratio	$2/3$	-
α_{slot}	ratio of slot opening to segments distance	0.7	-
a_{path}	number of parallel branches of each phase	6	-
B_{coil}	width of the coil around a segment	4.14	mm
w_{coil}	number of turns per coil around a segment	26	-

Table 5.3. Parameters of the basic electromagnetic design of PMAFM-ISA

5.3 Further optimization of PMAFM-ISA with 3D FEM

In this section, the basic electromagnetic design should be further optimized with the developed 3D FEM simulation platform described in the chapter 4. Because the limiting characteristic curve or speed-torque-curve is closely associated with the number of turns in a coil around each segment, it is assumed that the number of turens can vary continuously to achieve the same maximum torque 230 N m for a better comparison.

5.3.1 Width of coil B_{coil}

The width of coil B_{coil} illustrated in Fig. 5.5 is calculated based on the equivalent magnetic network of PMAFM-ISA and the specified maximum allowable flux density, which determines the area ratio between the copper coil and the SMC stator bar.

The PMAFM-ISA with different B_{coil} are listed in Table 5.4, while the limiting characteristic curves are illustrated in Fig. 5.12. Considering the magnetization curve of the SMC Somaloy 700HR3P illustrated in Fig. 3.11 and the necessary mechanical strength, the B_{coil} is adjusted to 3.82 mm.

Version	Coil width	Flux density at no-load
Basic design	4.14 mm	1.39 T
PMAFM-ISA 1	3.82 mm	1.33 T
PMAFM-ISA 2	3.47 mm	1.27 T
PMAFM-ISA 3	3.09 mm	1.21 T

Table 5.4. PMAFM-ISA design with different coil width B_{coil}

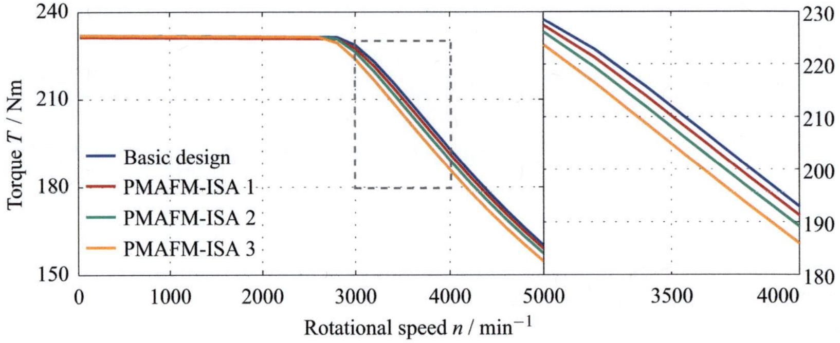

Figure 5.12. Limiting characteristic curves of PMAFM-ISA designs with different B_{coil}

5.3.2 Skewing of PM

A significant disadvantage of the DLCW with elementary winding 2A1 is the large torque ripple. In order to reduce the torque ripple of PMAFM-ISA 1, the skewing of PM is one of the most effective method. There are many different ways to skew PM and two of them are achieved by rotating the end edges of PM with the opposite and the same directions respectively, as illustrated in Fig. 5.13.

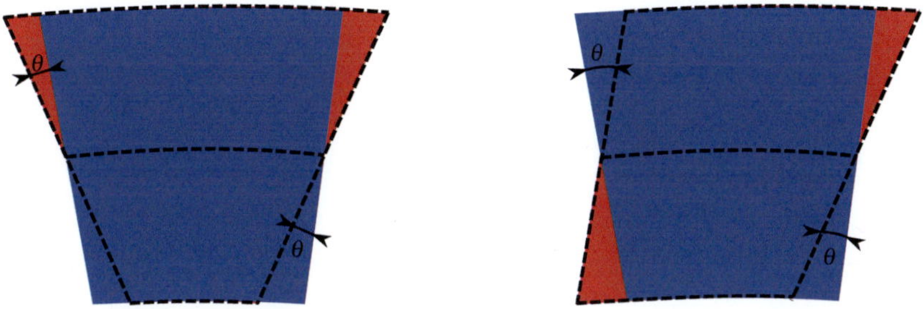

(a) Skewing method 1 (b) Skewing method 2

Figure 5.13. Different skewing methods of PM

When the PMAFM-ISA designs with differently skewed PM are fed with the maximum current, the torque ripples are listed in Table 5.5.

Version	**Skewing method**	**Skewing angle θ**	**Max. Torque ripple ΔT**
PMAFM-ISA 1	-	-	54.18 N m
PMAFM-ISA 4	1	12°	16.65 N m
PMAFM-ISA 5	1	15°	13.62 N m
PMAFM-ISA 6	2	15°	14.44 N m
PMAFM-ISA 7	2	18°	10.78 N m

Table 5.5. PMAFM-ISA designs with differently skewed PM

It should be noted that the skewing angle θ for the first method should be limited to avoid the overlapping of the adjacent PMs. In order to achieve an smaller torque ripple less than 11 N m, the PM is skewed with the second method and the skewing angle equals 18°. However, it should be noted that a larger skewing angle leads to considerable deterioration of the electromagnetic properties of PMAFM-ISA. Therefore, a larger skewing angle is not regarded any more.

5.3.3 Slot opening factor α_{slot}

Subsequently, the PMAFM-ISA design 7 is further optimized by varying the slot opening factor α_{slot} in terms of the limiting characteristic curves. The designs with different α_{slot} are listed in Table 5.6, while the limiting characteristic curves are illustrated in Fig. 5.14.

Version	**Slot opening factor**	**Max. Torque ripple ΔT**
PMAFM-ISA 7	0.7	10.78 Nm
PMAFM-ISA 8	1.0	13.85 Nm
PMAFM-ISA 9	0.4	17.23 Nm

Table 5.6. PMAFM-ISA designs with different slot opening factor α_{slot}

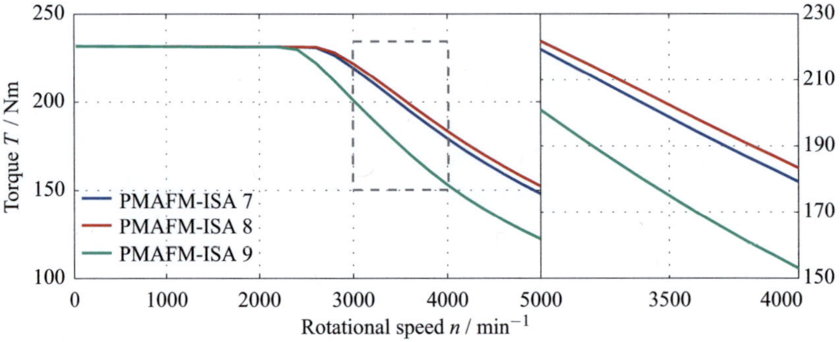

Figure 5.14. Limiting characteristic curves of PMAFM-ISA designs with different α_{slot}

It can be noted that the limiting characteristic curve of PMAFM-ISA rises with higher slot opening factor. In addition, a slot opening factor equal to 1 implies there is no lateral extrusion of the stator shoe, which simplifies the manufacturing process considering the brittle mechanical property of SMC. Due to the above consideration, the slot opening factor α_{slot} is determined as 1. On the other side, the disadvantage, namely the slightly increased torque ripple, has to be accepted.

5.3.4 Pole arc to pole pitch ratio α_{pm}

Based on the specified geometrical parameters, the PM form including the thickness L_{pm} and the ratio of PM arc to pole pitch α_{pm} are varied. Hereby, the PM mass should remain unchanged. The investigated PMAFM-ISA designs with different PM forms and the torque ripple are listed in Table 5.7, while the limiting characteristic curves are illustrated in Fig. 5.15.

Version	**PM Thickness**	**Ratio α_{pm}**	**Max. Torque ripple ΔT**
PMAFM-ISA 8	2.2 mm	2/3	13.85 Nm
PMAFM-ISA 10	1.96 mm	0.75	14.38 Nm
PMAFM-ISA 12	2.5 mm	0.59	17.82 Nm

Table 5.7. PMAFM-ISA designs with different pole arc to pole pitch ratio α_{pm}

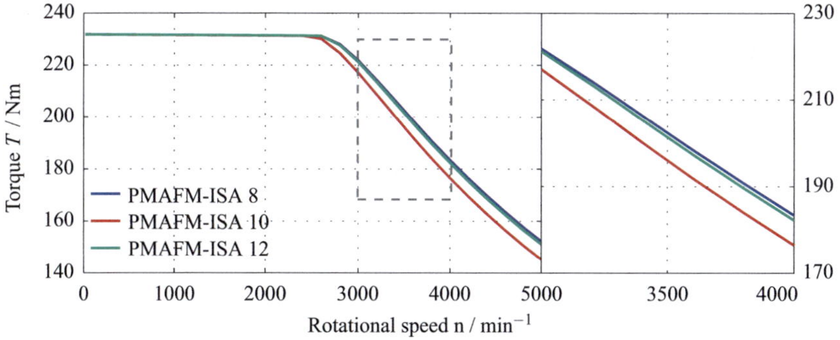

Figure 5.15. Limiting characteristic curves of PMAFM-ISA depending on α_{pm}

From the above 3D FEM results, it is clear that a broad and thin PM leads to improvement of the limiting characteristic curve. In comparison, a thicker and narrower PM results in higher torque ripple. Based on the above analysis, the PM form remains unchanged. It should be noted that the PM demagnetization of the ultimate PMAFM-ISA design must be carefully examined due to the small PM thickness, especially at high temperature.

5.3.5 Thickness of stator shoe L_{shoe}

Thickness of stator shoe is another important geometric parameter. The PMAFM-ISA designs with different thickness of stator shoe are listed in Table 5.8 and the limiting characteristic curves are illustrated in Fig. 5.16.

Version	Stator shoe thickness
PMAFM-ISA 8	5 mm
PMAFM-ISA 13	0 mm
PMAFM-ISA 14	2 mm
PMAFM-ISA 15	8 mm

Table 5.8. PMAFM-ISA designs with different thickness of stator shoe L_{shoe}

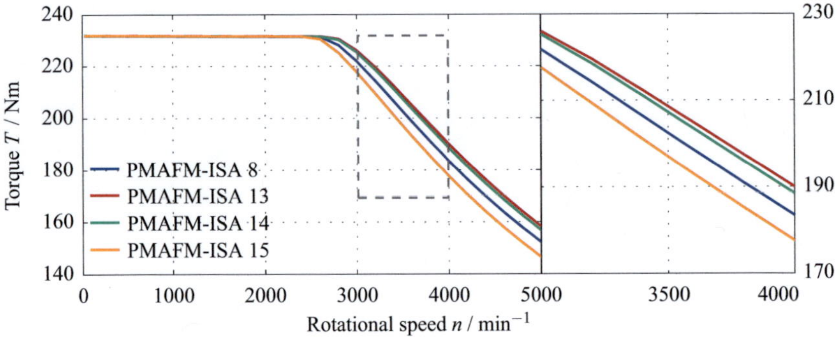

Figure 5.16. Limiting characteristic curves of PMAFM-ISA depending on L_{shoe}

It can be seen that the PMAFM-ISA performance becomes better with decreasing thickness of stator shoe. However, the thickness of the stator shoe L_{shoe} remains unchanged considering that a stator shoe is extremely helpful in the winding process and a minimum thickness of 5 mm is necessary due to the brittle mechanical property of SMC.

5.3.6 PM form

For the current optimal design PMAFM-ISA 8, there is an extremely small arc at the inner radius of PM, which can result in great manufacturing difficulty. For this reason, the form is slightly changed by cutting away the PM tips to achieve a flat bottom, as illustrated in Fig. 5.17.

Figure 5.17. Changing of PM form to achieve a flat bottom

Because the PM mass should be kept as constant, the PM thickness L_{pm} or the pole arc to pole pitch ratio α_{pm} need to be changed. The PMAFM-ISA designs with the same PM mass and different PM form are listed in Table 5.9, while the limiting characteristic curves are illustrated in Fig. 5.18.

Version	**PM Thickness**	**Ratio α_{pm}**	**Max. Torque ripple ΔT**
PMAFM-ISA 8	2.2 mm	2/3	13.85 N m
PMAFM-ISA 16	2.3 mm	2/3	11.41 N m
PMAFM-ISA 17	2.2 mm	0.7	10.46 N m

Table 5.9. PMAFM-ISA designs with different PM forms

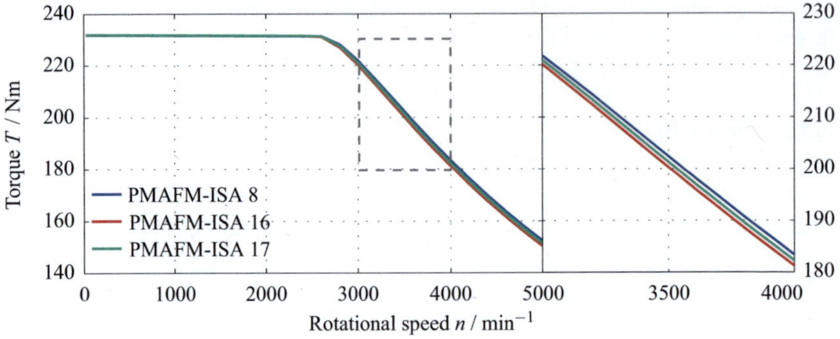

Figure 5.18. Limiting characteristic curves of PMAFM-ISA depending on PM forms

It can be noted that the PMAFM-ISA 17 with broader PM has better performance by observing the limiting characteristic curve. In addition, the torque ripple is further reduced to less than 11 N m. As a result, the PMAFM-ISA design 17 is considered as the most promising design temporarily.

5.3.7 SMC material

For the above optimization, the SMC material Somaloy 700HR3P is used to for the stator segments and rotor yoke. However, as mentioned in chapter 3, the components manufactured from this material cannot be machined because of the lack of binder material. For the ideal case, the necessary components should be directly pressed. Due to the unacceptable high costs of press system, the stator segments and rotor yoke for the prototype should be milled from the standard cylinder. For this reason, it is necessary to compare the PMAFM-ISA designs with components made from the other manufacturable SMC materials. The designs with different SMC material are listed in Table 5.10 and their limiting characteristic curves are illustrated in Fig. 5.19.

It can be noted the performance of the PMAFM-ISA design 17 declines slightly as the SMC material with binder material shows higher losses and lower permeability. In addition, there is no obvious difference between the PMAFM-ISA made from the SMC Siron®S300b and Siron®S400b.

Version	Utilized SMC Material
PMAFM-ISA 17	Somaloy 700HR3P
PMAFM-ISA 18	Siron®S300b
PMAFM-ISA 19	Siron®S400b
PMAFM-ISA 20	Siron®STestb

Table 5.10. PMAFM-ISA designs with different SMC material

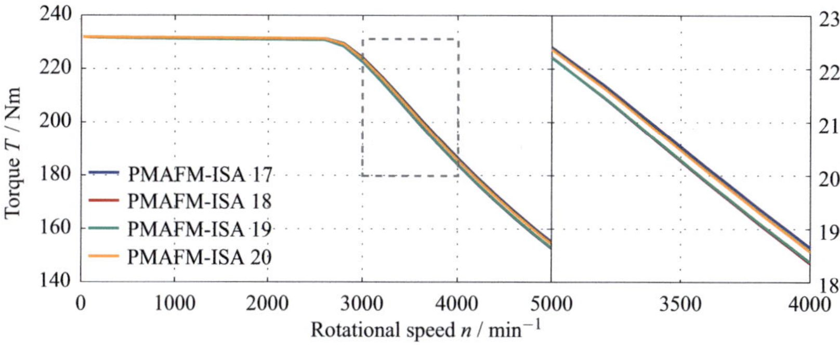

Figure 5.19. Limiting characteristic curves of PMAFM-ISA depending on SMC material

In comparison, the PMAFM-ISA design based on the material Siron®STestb has better performance in terms of the limiting characteristic curves. Therefore, the PMAFM-ISA design 20 is determined temporarily as the best design.

5.3.8 Segmentation of PM

Eddy current losses in PM can significantly influence performance of electrical machine and are one of the most important thermal sources of PMAFM-ISA, where PMs are mounted on the surface of the rotor yoke [S7]. One of the most effective methods to reduce the PM eddy current losses is the segmentation of PM, which can be either segmented along the radial direction, the circumferential direction or a combination of both directions, namely the 2D segmentation.

The eddy current losses in all PMs of PMAFM-ISA designs with differently segmentation of PM at no-load and 60 min^{-1} are listed in Table 5.11. It is apparent that the PM segmentation along the radial direction leads to a more even area of PM segments and can achieve less losses. Compared with the direction of segmentation, the number of PM segments or rather the area of per segment is more dominant. For instance, a PM consisting of 6 segments can generate nearly only half of the PM eddy current losses compared with PMAFM-ISA design 20. The informative efficiency map of PMAFM-ISA 25 is illustrated in Fig. 5.20.

Version	**Segmentation method**	**Number of segments**	**PM losses**
PMAFM-ISA 20	-	-	15.65 W
PMAFM-ISA 21	circumferential	2	12.17 W
PMAFM-ISA 22	circumferential	3	9.46 W
PMAFM-ISA 23	radial	2	10.70 W
PMAFM-ISA 24	2D	4	8.84 W
PMAFM-ISA 25	2D	6	7.28 W

Table 5.11. PMAFM-ISA designs with different PM segmentation

With the help of PM segmentation, the efficiency of PMAFM-ISA 25 is obviously improved and the peak efficiency increases more than 1.5 % compared with PMAFM-ISA 20. In addition, the reduction of the peak torque at low speed area becomes much smaller and the speed-torque-curve becomes much flatter.

5.4 Final electromagnetic design and validation

In spite of the great advantage of PM segmentation, the idea is abandoned for the final design of the first prototype to reduce the assemble difficulty and to improve the mechanical firmness of PM. The final electromagnetic design is obtained based on the above PMAFM-ISA 20 in terms of the following aspects:

- Segmenting the rotor yoke into 12 parts due to the limited available size of pressed standard SMC cylinder.
- Design of small gaps on the rotor yoke for the adhesive, which aims to glue the PM with the rotor yoke.

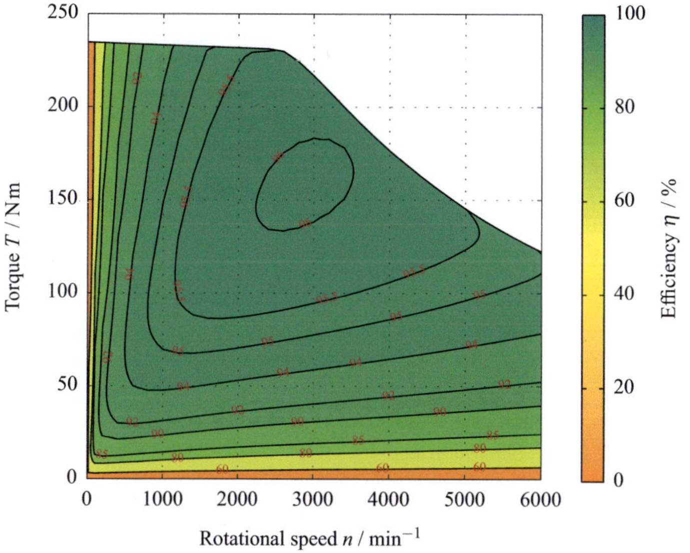

Figure 5.20. Efficiency map of PMAFM-ISA 25 with 2D PM-segmentation

- Rounding of SMC components to reduce the stress concentration resulted from manufacturing and to avoid damage of winding isolation.

The ultimately manufactured stator segment and rotor yoke element are illustrated in Fig. 5.21.

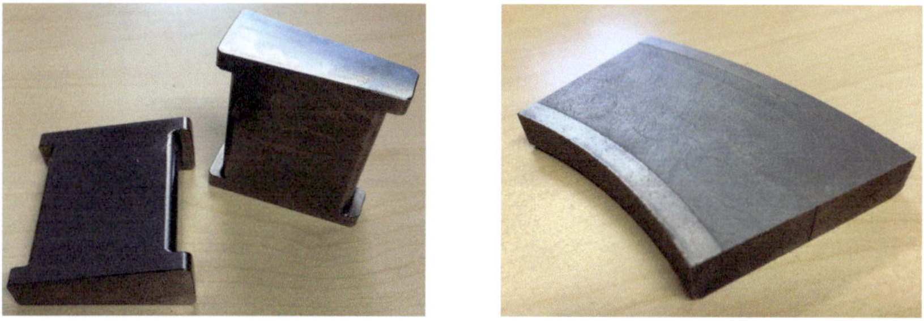

(a) Stator segments

(b) Rotor yoke component

Figure 5.21. Manufactured SMC parts

It is apparent that the inner and outer edges of the rotor yoke are 0.1 mm higher than the rest part to realize the necessary gap for the adhesive.

Although the cross sectional area is reduced due to the rounding of the stator segments and the reluctance becomes larger due to the adhesive gap between PM and rotor yoke, their influence on the electromagnetic properties of PMAFM-ISA is extremely limited. The efficiency map of the final electromagnetic design is illustrated in Fig. 5.22.

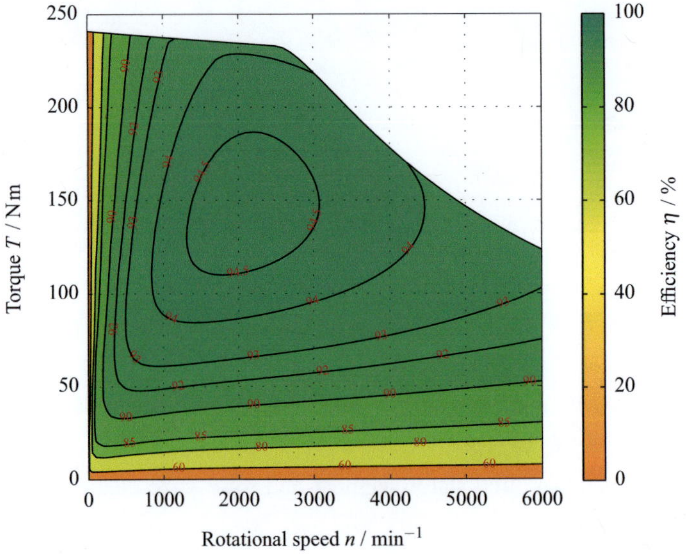

Figure 5.22. Efficiency map of the final design of PMAFM-ISA without PM-segmentation

The PM has the relatively low thickness equal to 2.2 mm and its demagnetization should be carefully investigated. As shown in Fig. 5.23, the working point of PM at 160 °C and I_d equalling 640 A, which is 1.5 times of the maximum phase current, is marked with a small red circle. It can be seen that even for the worst case, the flux density of PM is high enough and the demagnetization doesn't occur.

At last, the most important parameters of the final electromagnetic design of PMAFM-ISA are summarized in Table 5.12.

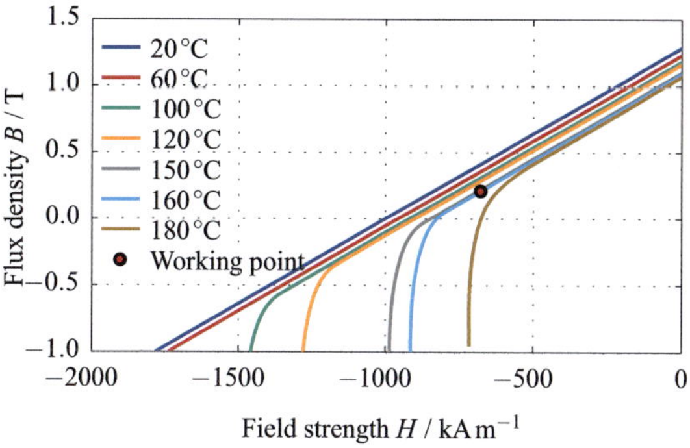

Figure 5.23. Demagnetization test of PM in final design

Symbol	Description	Value	Unit
D_a	active outer diameter	301	mm
D_i	active inner diameter	209	mm
L_{motor}	active total axial length	75	mm
$N_{segments}$	Number of stator segments	36	-
p	Number of pole pairs	12	-
L_g	axial length of air gap	1	mm
L_{shoe}	axial length of stator shoe	5	mm
L_{bar}	axial length of stator bar	39.6	mm
L_{yoke}	axial length of rotor yoke	9.5	mm
N_{yoke}	Number of segments per rotor yoke	9	-
L_{PM}	axial length of PM	2.2	mm
α_{slot}	ratio of slot opening to segments distance	0.7	-
α_{PM}	pole arc to pole pitch ratio	0.7	mm
θ_{PM}	skewing angle of PM	18	°
N_{phase}	Number of parallel branches of each phase	6	-
B_{coil}	width of the coil around a segment	3.80	mm
w_{coil}	Number of turns per coil around a segment	26	-

Table 5.12. Main parameters of the electromagnetic design of PMAFM-ISA

6

Electromagnetic Design of PMTFM-CP

Besides the PMAFM-ISA, the permanent magnet excited transverse flux machine with claw poles (PMTFM-CP) is another promising topology for the application of SMC, which is distinguished by its high pole pair number and torque density [O6, O7, S8]. In PMTFM-CP, the permanent magnets with opposite circumferential magnetization directions are embedded in the rotor. All the stator segments are located on one side of the rotor and a ring winding is located within the stator segments for each phase. Therefore, the PMTFM-CP can achieve high torque density and relatively simple mechanical construction simultaneously [S9, S10].

Assuming the phases of PMTFM-CP are magnetic decoupled from each other, it is only necessary to analyse one pole pair model of one phase considering the magnetic symmetry around the rotational axis. For instance, the meshed model and the FEM results at no-load are illustrated in Fig. 6.1. It can be clearly seen the magnetic flux flows three-dimensionally in the PMTFM-CP. The main magnetic flux marked with small arrows from the PM is concentrated in the rotor concentrator and then flows through the air gap, the stator claws, the teeth on the front side, the stator yoke, the teeth, the claws, and the air gap on the rear side, which finally flows back to the PMs in the rotor. To realize the conduction of 3D magnetic flux, the SMC material is used to manufacture the complex stator and the rotor flux concentrator.

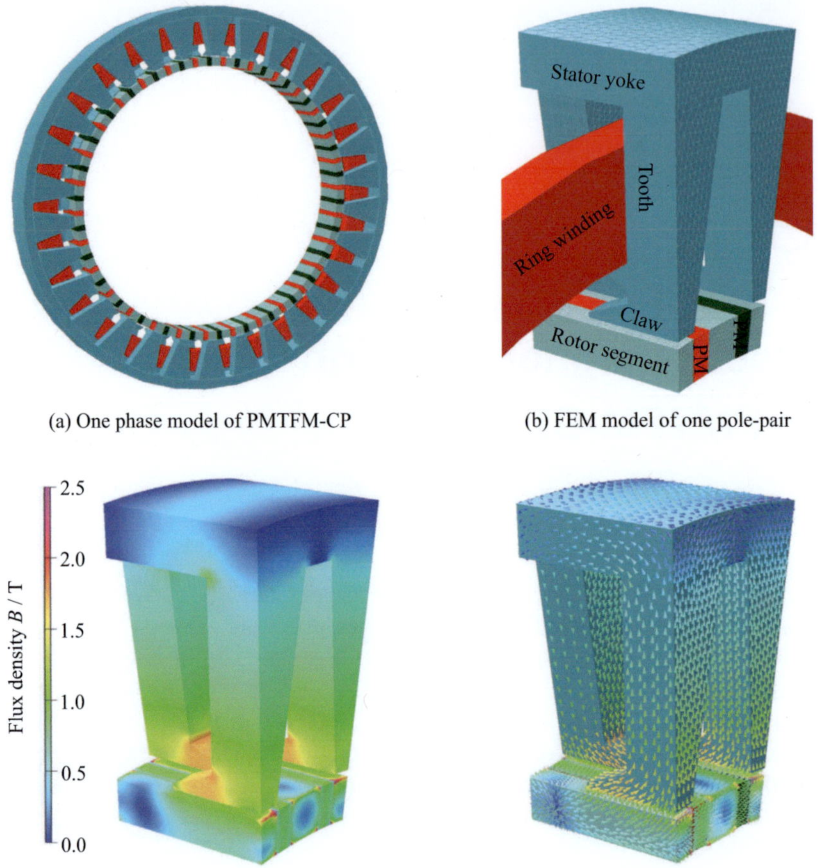

(a) One phase model of PMTFM-CP

(b) FEM model of one pole-pair

(c) *B*-Field

(d) Magnetic flux

Figure 6.1. Structure and FEM results of a PMTFM-CP

The chapter is organized as follows. At first, the free geometric parameters of PMTFM-CP, the optimization objectives and design procedure are presented. Subsequently, the electromagnetic design of PMTFM-CP has been carried out according to the design procedure.

6.1 Free design parameters and design objectives

In this section, the free design geometric parameters and the optimization objectives are introduced. Subsequently, the design procedure is described, which is based on the standard components including Taguchi method, the nondominated sorting genetic algorithm 2 (NSGA2), the multi-objective particle swarm optimization (MOPSO) and the fuzzy logic (FL).

6.1.1 Free design parameters of PMTFM-CP

Because of the complex structure of the stator segments, many basic variables are needed to completely determine the geometry of PMTFM-CP. Some of them can be clearly found in the side view, as shown in Fig. 6.2.

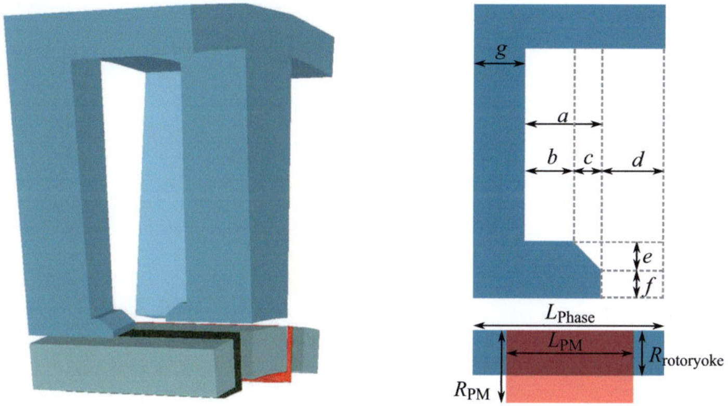

Figure 6.2. Side view of PMTFM-CP

With help of these variables, the axial length and radial length of all components can be determined. The other necessary basic geometric variables to determine the tooth and claw are defined in the front view. As illustrated in Fig. 6.3, the modelling process consists of three main steps, which are the basic modelling, the slight modification of stator tooth and claw respectively,

The angles θ_1 and θ_2 are the angles of the SMC concentrator in the rotor and the basic angle of the stator tooth and claw. In the following step, the basic geometry of the stator can be modified with the help of the angles θ_3 and θ_4, which are used to broaden the stator tooth and claw respectively. The angle $\theta_{3\max}$ is the maximum allowable deflection angle of stator tooth, for which the two adjacent stator teeth contact each other. The angle $\theta_{4\max}$ corresponds to the angle, for which the two edges of the stator claw are parallel with each other.

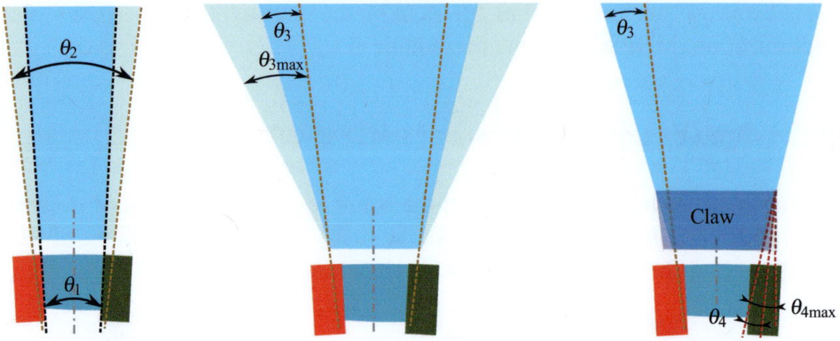

(a) Basic modelling (b) Modification of stator tooth (c) Modification of claw

Figure 6.3. Front view of modelling procedure of PMTFM-CP

Based on the above basic geometric variables, totally 10 free parameters are defined and listed in Table 6.1. It is noted that most of them are defined as ratio to achieve a general valid conclusion.

6.1.2 Design objectives

Due to the time-consuming 3D FEM, the PMTFM-CP is optimized only for one working point with help of the stationary FEM, namely the PMTFM-CP is fed with the maximum phase current along the q-axis. Considering the requirements of the application in the electric vehicle, five objectives are defined in this work.

With help of the 3D FEM, the flux linkage Ψ_u of the phase u, the inner torque T for different electrical rotor positions θ can be determined. On this basis, the torque and flux linkage of the other two phases Ψ_v and Ψ_w can be obtained by translating Ψ_u for the electrical angle 120° and 240° respectively. After then, the flux linkage Ψ_d along the direct axis and Ψ_q along the q-axis can be calculated:

Parameter	Description	Definition
L_{tooth}	axial length of the stator tooth	g
L_{claw}	axial length control of the stator claw	$d/(a+d)$
L_{short}	axial length adjustment of bevel at claw tip	$c/(b+c)$
R_{short}	axial length adjustment of bevel at claw tip	$e/(e+f)$
W_{basic}	basic angle control of the stator tooth and claw	θ_2/θ_1
W_{tooth}	angle adjustment of the stator tooth	θ_3/θ_{3max}
W_{claw}	angle adjustment of the stator claw	θ_4/θ_{4max}
L_{PMR}	ratio of the axial length of PM to one phase	L_{PM}/L_{phase}
R_{RYR}	ratio of the radial length of SMC concentrator to PM	R_{yoke}/R_{PM}
p	the pole-pair number of PMTFM-CP	p

Table 6.1. Free design parameters of PMTFM-CP

$$\Psi_d = \frac{2}{3}\left(\Psi_u \cos(\theta) + \Psi_v \cos(\theta - 120°) + \Psi_w \cos(\theta - 240°)\right) \quad (6.1)$$

$$\Psi_q = \frac{2}{3}\left(\Psi_u \cos(\theta + 90°) + \Psi_v \cos(\theta - 30°) + \Psi_w \cos(\theta - 150°)\right) \quad (6.2)$$

Subsequently, the stationary voltages along the d- and q-axis u_d and u_q can be calculated, where ω_{el} is the electrical rotational speed of PMTFM-CP and R_S is the resistance of one phase.

$$u_d = R_S \cdot i_d - \omega_{el} \cdot \Psi_q \quad (6.3)$$

$$u_q = R_S \cdot i_q + \omega_{el} \cdot \Psi_d \quad (6.4)$$

As the first optimization objective, the characteristic speed ω_{chara} is defined as the mechanical speed, at which the voltage limit is achieved. It can be determined through solving the (6.5), where u_{max} is the amplitude of the maximum allowable phase voltage.

$$u_d^2 + u_d^2 = u_{max}^2 \quad (6.5)$$

Correspondingly, the ω_{chara} can be calculated using (6.6).

$$\omega_{\text{chara}} = \frac{\omega_{\text{el}}}{p \cdot 2 \cdot \pi} \tag{6.6}$$

The second objective is the copper losses, which is one of the dominant losses when the PMTFM-CP is fed with the maximum phase current. It can be calculated with (6.7), where i_{max} is the maximum effective phase current.

$$P_{\text{cu}} = i_{\text{max}}^2 \cdot R_{\text{S}} \tag{6.7}$$

The corresponding output torque at the characteristic speed ω_{chara} is the third objective and can be calculated using (6.8), where p_{fe} is equal to the iron losses of all SMC parts at the speed ω_{chara}.

$$T_{\text{max}} = T - \frac{p_{\text{fe}}}{\omega_{\text{chara}} \cdot 2 \cdot \pi} \tag{6.8}$$

The fourth objective is the efficiency η_{chara} of PMTFM-CP at the characteristic speed η_{chara} calculated with (6.9). Hereby, the friction losses and the eddy current losses in PMs are not taken into account.

$$\eta_{\text{chara}} = \frac{T_{\text{max}} \cdot \omega_{\text{chara}}}{T_{\text{max}} \cdot \omega_{\text{chara}} + p_{\text{fe}} + p_{\text{cu}}} \tag{6.9}$$

The last objective is the torque ripple ratio α_{T} determined using (6.10), which is defined as the ratio between the torque ripple to the average output torque T_{max}.

$$\alpha_{\text{T}} = \frac{T(\theta)_{\text{max}} - T(\theta)_{\text{min}}}{T_{\text{max}}} \tag{6.10}$$

All of the above mentioned five design objectives are summarized and listed in Table 6.2. Besides the predefined geometric constraints, the properties of the inverters and the required PM mass listed in Table 1.2, there are some other conditions applied in the design process. First of all, the electrical machine is only fed with the current component i_{q} along the q-axis considering the small difference between the inductances along the d- and q-axis. Besides, the number of coils per phase is determined as 5 to achieve an appropriate compromise among the objectives. At last, the ratio between the angle of PM to the concentrator remains constant equal to $1/2$.

Objective	Unit	Description
ω_{chara}	min^{-1}	characteristic speed of PMTFM-CP
P_{cu}	W	copper losses
T_{max}	N m	maximum torque at ω_{chara}
η_{chara}	%	efficiency of PMTFM-CP at ω_{chara}
α_{T}	%	torque ripple ratio

Table 6.2. Design objectives of PMTFM-CP

6.1.3 Design procedure of PMTFM-CP

The design procedure of PMTFM-CP in this work is illustrated in Fig. 6.4, which is based on a combination of many standard methods. Because the 3D FEM is extremely time-consuming, it is impossible to optimize all the 10 free parameters of PMTFM-CP listed in Table 6.1. Therefore, a design of experiment method (DoE), the Taguchi method, is firstly applied to evaluate the importance of each design parameter in consideration of the defined five objectives. Subsequently, the selected critical parameters are further optimized with the nondominated sorting genetic algorithm 2 (NSGA2) and the multi-objective particle swarm optimization (MOPSO) [O8, O9, S11].

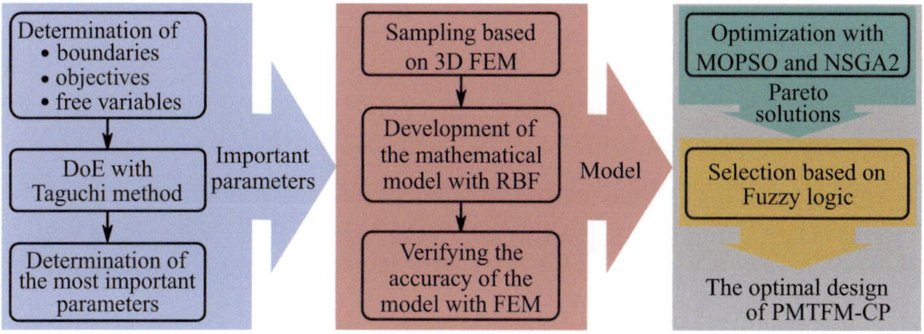

Figure 6.4. Design process of PMTFM-CP

The genetic algorithms (GA) are based on an analogy to the evolution process in nature and often adopted to optimize electrical machines. Among many different GA, the newly developed NSGA2 is distinguished by its effectiveness and reliability, which is suitable for the multi-objective optimization. In comparison, the PSO is a heuristic search technique and simulates the social behaviour of a flock of birds searching food. The MOPSO is an extension of PSO, which aims to solve the multi-objective optimization problem. Both methods have been widely used in many different cases.

Because for both algorithms a large number of PMTFM-CP designs should be analysed, a mathematical model based on the radial basis function (RBF) is developed and verified with FEM in this work, so that the five objectives depending on the design parameters can be analytically fast calculated. Since the five objectives are normally in conflict with each other, there is no explicit design with the best values for all objectives. Therefore, the results of both optimization methods are a lot of outstanding "trade-off" designs, which are called nondominated solutions or Pareto front representing the best possible comprises among the objectives. Afterwards, the optimal design is selected from the Pareto front with the help of fuzzy logic (FL). Finally, the optimal design is validated with the 3D FEM.

6.2 Pre-optimization with Taguchi method

Depending on different values of the 10 free parameters in Table 6.1, numerous designs of PMAFM-ISA can be developed. Because it is too time-consuming to optimize all parameters and the influence of these parameters on the 5 objectives are distinctly different, the Taguchi method is utilized to determine the importance of the design parameters with respect to the objectives [262–264]. Only the relatively more important parameters are selected and further optimized. Meanwhile, the values of other parameters in the design space can be determined.

The Taguchi method is based on an orthogonal array (OA), which should be determined depending on the number and levels of parameters to be investigated. In the OA, the different combinations of the values of the parameters are specified. With the help of Taguchi method, a large number of parameters can be researched with a small number of PMTFM-CP designs.

Considering the free parameters listed in Table 6.1, the OA $L_{32}(2^1 \times 4^9)$ is utilized in this thesis, which implies there are 9 parameters with 4 levels and 1 parameter with 2 levels. Besides, totally 32 experiments or rather designs of PMTFM-CP need to be investigated. A possible value or rather a level of a parameter P is defined as a factor P_i and all factors are listed in Table 6.3.

Parameter		Level1	Level2	Level3	Leve4
$A:$	W_{basic}	1	1.2	-	-
$B:$	L_{tooth}	8	9	10	11
$C:$	W_{tooth}	0	1/3	2/3	1
$D:$	W_{claw}	0	1/3	2/3	1
$E:$	L_{claw}	0.25	0.5	0.75	1
$F:$	L_{short}	0.2	0.4	0.6	0.8
$G:$	R_{short}	0.2	0.4	0.6	0.8
$H:$	R_{RYR}	0.7	0.8	0.9	1
$I:$	L_{PMR}	0.7	0.8	0.9	1
$J:$	p	15	20	25	30

Table 6.3. Factors of all free design parameters

According to the OA $L_{32}(2^1 \times 4^9)$, 32 PMTFM-CP designs should be analysed, which correspond the different combinations of the above defined factors. Some of them are listed in Table 6.4.

Design	Parameter									
	A	B	C	D	E	F	G	H	I	J
D_1	A_1	B_1	C_1	D_1	E_1	F_1	G_1	H_1	I_1	J_1
D_2	A_1	B_1	C_2	D_2	E_2	F_2	G_2	H_2	I_2	J_2
D_3	A_1	B_1	C_3	D_3	E_3	F_3	G_3	H_3	I_3	J_3
					⋮					
D_{30}	A_2	B_4	C_2	D_4	E_3	F_1	G_3	H_1	I_2	J_4
D_{31}	A_2	B_4	C_3	D_1	E_2	F_4	G_2	H_4	I_3	J_1
D_{32}	A_2	B_4	C_4	D_2	E_1	F_3	G_1	H_3	I_4	J_2

Table 6.4. Orthogonal array $L_{32}(2^1 \times 4^9)$

Based on the above two Tables, totally 32 different PMTFM-CP designs have been developed and analysed with the 3D FEM. Subsequently, the five objectives of each design and the mean value are determined are listed in Table 6.5.

Design	T_{max} [Nm]	J_{slot} [A/mm^2]	ω_{chara} [min^{-1}]	η_{chara} [%]	α_T [%]
D_1	92.65	9.57	3208	94.68	39.27
D_2	127.36	9.12	2099	94.27	24.03
D_3	156.43	8.54	1679	94.92	8.83
		⋮			
D_{30}	179.78	26.75	926	91.56	9.73
D_{31}	124.73	33.81	2055	94.17	24.18
D_{32}	144.87	33.81	1189	91.5	25.5
Mean	132.6	17.71	1650	92.01	21

Table 6.5. Calculated values of the five objectives of 32 PMTFM-CP designs

Based on results in Table 6.5, the influence of each parameter on every objective can be determined through the variance analysis (ANOVA). For instance, the relative importance of the parameter in consideration of the objective T_{max} is analysed through the ANOVA.

At first, the mean effect of each factor P_i on the objective T_{max} can be calculated with (6.11), where m is the number of the total designs D_i including the factor P_i. The determined mean effect of all factors on the objective T_{max} is illustrated in Fig. 6.5.

$$\bar{T}_{max}(P_i) = \frac{1}{m} \sum_{i=1}^{m} T_{max}\,(\,D_i\,) \tag{6.11}$$

Subsequently, the sum of the squares (SSF) of the differences between the mean effect of all n factors belonging to the parameter P and the mean value of T_{max} from Table 6.5 can be computed with (6.12).

$$SSF(P) = \frac{32}{n} \sum_{i=1}^{n} (\,\bar{T}_{max}\,(\,P_i\,) \;-\; mean(\,T_{max}\,)\,)^2 \tag{6.12}$$

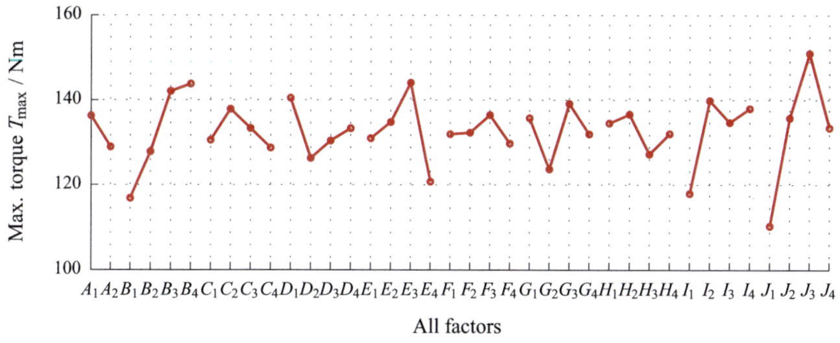

Figure 6.5. Mean effect of each factor on the objective T_{max}

Finally, the influence of the parameter P on the objective T_{max} can be calculated using (6.13).

$$P_{Inf} = \frac{SSF(P)}{SSF(A) + SSF(B) + \cdots + SSF(J)} \tag{6.13}$$

Similar to the objective T_{max}, the influence of all 10 parameters on five objectives can be calculated, as illustrated in Fig. 6.6.

Based on the influence of each parameter on the objectives, the most critical 4 parameters and their optimization range are determined. In addition, the values of other parameters are determined as well considering the mean effect of all factors of each parameter. The results of Taguchi method are listed in Table 6.6.

Parameter	**A** W_{basic}	**B** L_{tooth}	**C** W_{tooth}	**D** W_{claw}	**E** L_{claw}
Result	1	$[8 \sim 10]$	0	0	$[0.5 \sim 0.9]$

Parameter	**F** L_{short}	**G** R_{short}	**H** R_{RYR}	**I** L_{PMR}	**J** p
Result	0.6	$[0.6 \sim 0.8]$	0.8	$[0.7 \sim 0.9]$	20

Table 6.6. Results of pre-optimization based on Taguchi method

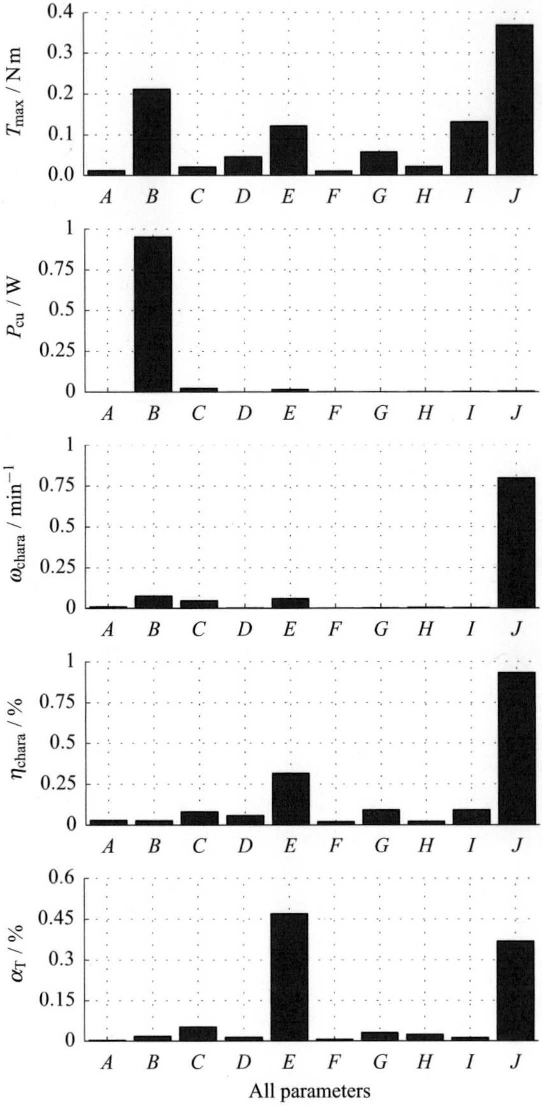

Figure 6.6. Influence of all parameters on the five objectives

It can be noted the pole-pair number has dominant influence on 4 objectives, which is determined as 20 in order to achieve an appropriate comprise among the objectives as well as the requirements on the inverter.

6.3 Mathematical model based on RBF

Afterwards, the 4 critical parameters should be further optimized with the algorithms non dominated sorting genetic algorithm 2 (NSGA2) and multi-objective particle swarm optimization (MOPSO). Considering the large computational expense of both algorithms, a surrogate mathematical model is needed to calculate the five objectives depending on the following design vector analytically and accurately.

$$\left[\begin{array}{cccc} L_{\text{tooth}} & W_{\text{claw}} & R_{\text{short}} & L_{\text{PMR}} \end{array}\right]$$

In the past time, many different models have been developed and the mathematical model based on the radial basis function (RBF) is used in this work [265–267]. For this purpose, 2025 designs of PMTFM-CP corresponding to a range of the design vector have been analysed with 3D FEM, which can be considered as 2025 sampling points distributed evenly in the design space. On this basis, the mathematical model including five equations is developed, which is described in (6.14).

$$f(x) = \beta \cdot x + \sum_{k=1}^{n} \lambda_{\text{k}} \cdot \Phi_{\text{k}}(x) \tag{6.14}$$

Hereby, the $f(x)$ is one of the five objectives, x is the design vector, β is the first order polynomial factor and can be calculated with the least square method, n is the number of the sampling points, λ_k is the weighting coefficient of the k-th RBF Φ_k.

The RBF $\Phi(x)$ is a real value function, whose value is only depending on the Euclidean distance between the design vector x and the origin x_0. There are many different kinds of RFB, the Gaussian RBF expressed in (6.15) is utilized in this work, where θ is a prescribed positive constant named shape parameter, $D_{\max}$ is the maximum distance between any two of the sampling points.

$$\Phi_{\text{k}}(x) = \exp\left(-\theta \cdot \frac{\|x - x_{0\text{k}}\|^2}{D_{\max}}\right) \tag{6.15}$$

Before the developed mathematical model is applied to the optimization, its accuracy should be validated. For this purpose, another 91 testing points have been analysed with 3D FEM and the results are compared with that from the mathematical model. In order to achieve a persuasive conclusion, the testing points are determined with the Latin Hypercube Sampling (LHS) method, which is a statistical method used to generate a random and roughly even distribution of the sampling points. In addition, the identical values of sampling points in any dimension of the design space can be avoided.

In order to evaluate the precision of the model, the normalized root mean squared error (NRMSE) of all five objectives for the 91 testing points is calculated using (6.16),

$$\mathrm{NRMSE} = \sqrt{\frac{\sum_{k=1}^{91} \left(f_k\left(x_i\right) - g_k\left(x_i\right)\right)^2}{\sum_{k=1}^{91} f_k\left(x_i\right)^2}} \tag{6.16}$$

where $f_k(x_i)$ and $g_k(x_i)$ are the values of the k-th objective of the i-th sampling point calculated with the FEM and the mathematical model respectively.

The NRMSE of all objectives are calculated and listed in Table 6.7.It is apparent that the results from the mathematical model match well with that from 3D FEM. Thus, it can be concluded the model is precise enough and can be used to calculate the objective values instead of time-consuming 3D FEM in the following optimization. For a clear overview, the T_{max} depending on the two design param-

Objective	T_{max}	P_{cu}	ω_{chara}	η_{chara}	α_{T}
NRMSE in %	1.19	0.42	1.28	0.24	0.83

Table 6.7. NRMSE of all five objectives

eters L_{tooth} and L_{claw} are calculated with the developed mathematical model and compared with the FEM results, as shown in Fig. 6.7. It can be noted that the FEM results can be well approximated.

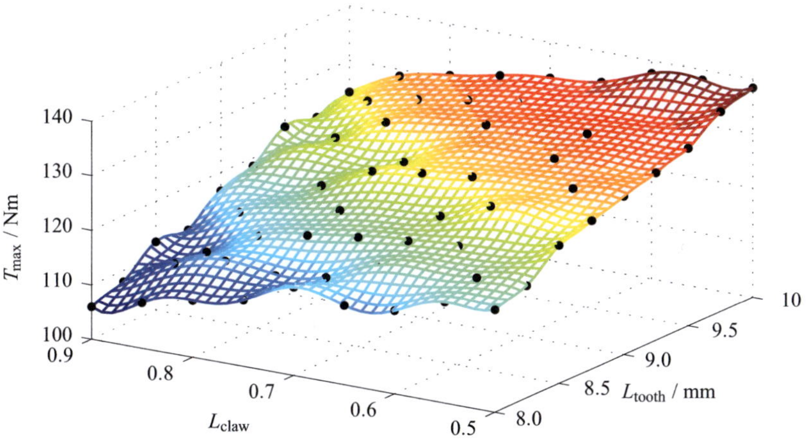

Figure 6.7. Comparison of the T_{max} calculated from the model and FEM

6.4 Optimization with NSGA2 and MOPSO

Because there are five objectives rather than only one to be optimized, there is normally no explicit PMTFM-CP design with the best values of all objectives. Instead, many outstanding designs exist, which can make good trade-off among these objectives. All these designs are named as Pareto solutions and form the Pareto front. For a Pareto solution, an improvement of any objective without worsening at least one of the other objectives is impossible. The purpose of this section is to find out the Pareto solutions with the MOPSO and NSGA2. The three important criteria used to evaluate the results of optimization algorithms are [268, 269]:

- The searched Pareto solutions should be located as closely as possible to the true Pareto front.
- The distribution of the searched Pareto solutions should be as even as possible.
- The determined Pareto solutions should cover the true Pareto front as widely as possible to obtain a good spread.

In order to evaluate the quality of the Pareto front and to compare the different algorithms, many performance indices have been suggested and compared [268]. However, it should be noted that the selection of an appropriate index is not a simple task. Furthermore, the index cannot always properly reflex the quality of the solution.

6.4.1 Optimization with MOPSO

The inspiration of PSO came from the cooperative behaviour of a flock of birds searching foods, which is developed by Kennedy and Eberhart in 1995. In this work, a PMTFM-CP design is considered as a particle, whose position corresponds the design parameters of PMTFM-CP and can be updated using (6.17),

$$x_i(t+1) = x_i(t) + v_i(t+1) \tag{6.17}$$

where $x_i(t)$ is the position of the i-th particle at the t-th iteration.

The velocity v_i of the i-th particle consists of three components: the own inertia, the own previous experience and the so-called "social part" of all particles, which can be calculated using (6.18).

$$\begin{aligned} v_i(t+1) = & w_i \cdot v_i(t) + c_1 \cdot r_1(t) \cdot (x_{\mathrm{i,pbest}}(t) - x_i(t)) + \\ & c_2 \cdot r_2(t) \cdot (x_{\mathrm{g,pbest}}(t) - x_i(t)) \end{aligned} \tag{6.18}$$

In (6.18), the r_1 and r_2 are random numbers between 0 and 1 to obtain the diversity of the population. The positive constants c_1 and c_2 are the cognitive and social factors, which are acceleration constants to control the convergence properties. The parameter w is called as the inertia weight factor aiming at controlling the impact of the previous velocity on the current velocity. Meanwhile, it can determine the trade-off between the global and local exploration abilities of the particle. The $x_{\mathrm{i,pbest}}(t)$ represent the best position previously attained by the i-th particle itself until t, while the $x_{\mathrm{g,pbest}}(t)$ is the global best position reached within the whole population as so far now.

The MOPSO is an extension of PSO aiming at multi-objective optimization, whose flow chart is illustrated in Fig. 6.8.

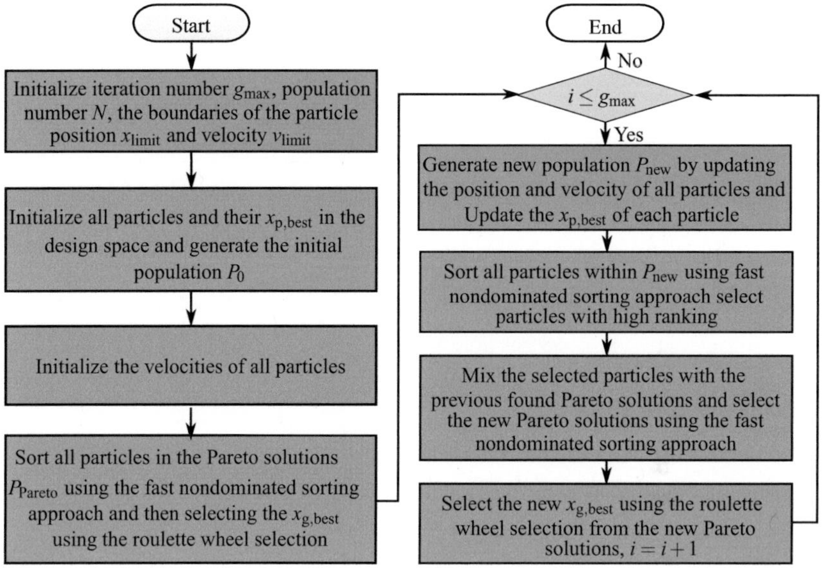

Figure 6.8. Flow chart of the MOPSO

The optimization has been performed for different combinations of population number N and iteration number g_{max} considering the randomness of the MOPSO [270, 271]. For a better overview, the initial particles and the searched Pareto solutions are plotted on the T_{max}-ω_{chara} plane. The results for 50 and 200 initial particles and the finally searched Pareto solutions are illustrated in Fig. 6.9. It should be noted that the searched Pareto solutions can also be projected to other planes of objectives and a projection is not a Pareto front in turn. Because there are five objectives to be optimized, the Pareto solutions in Fig. 6.9 cannot converge to an obvious curve as usual.

It can be seen that the initial particles are located in a broad region, which move into a relatively narrow region during the optimization procedure, so that either high T_{max}, high ω_{chara} or a good trade-off can be achieved.

Through the comparison of the above analysis results, it is clear that the number of the particles plays an important role. The more the initial particles are, the more Pareto solutions are found and the more even the distribution of the Pareto front is. Besides, there is no obvious change of the distribution of the Pareto front after certain iterations. Thus, the N and $g_{\max}$ should be appropriately set to avoid unnecessary calculation time and to well approximate the real Pareto front simultaneously.

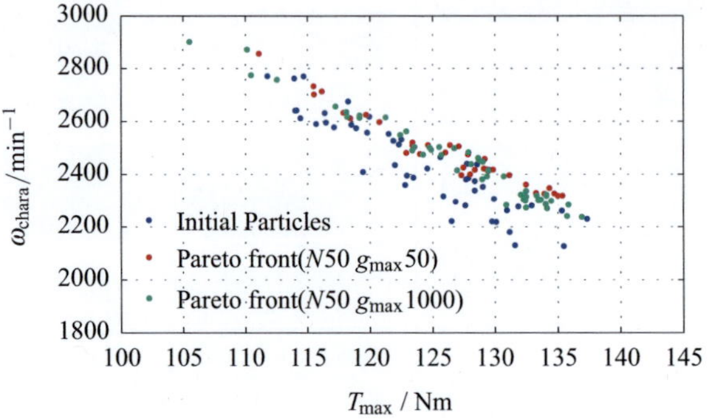

(a) 50 initial particles

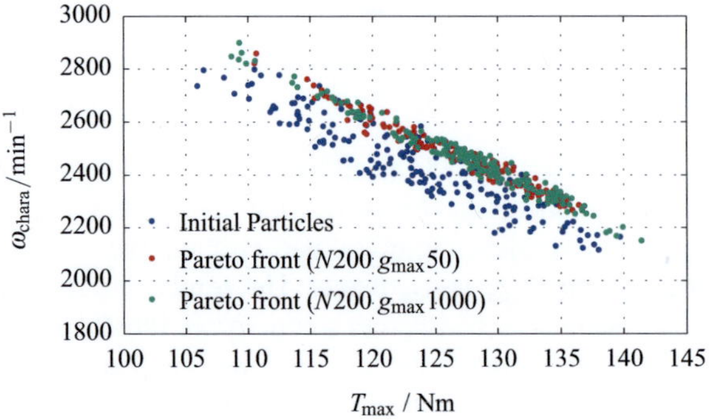

(b) 200 initial particles

Figure 6.9. Initial particles and searched Pareto solutions

6.4.2 Optimization with NSGA2

In comparison with the PSO, the optimization with GA is based on the simulation of the genetic replicated process. Hereby, each PMTFM-CP design corresponds to a chromosome and the design parameters are considered as the genetics of a chromosome. The flow chart of NSGA2 is shown in Fig. 6.10.

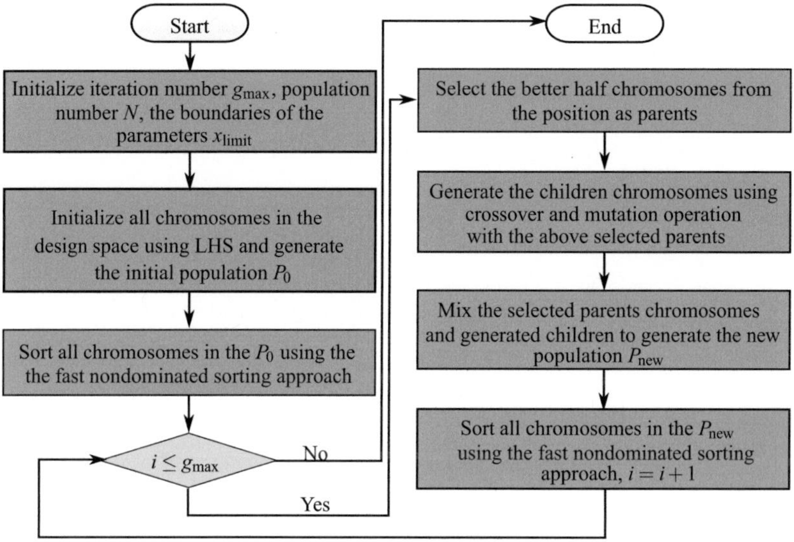

Figure 6.10. Flow chart of the NSGA2

For a better comparison, the optimization with NSGA2 has been performed for different combinations of N and g_{max} with the same initial designs as that of MOPSO. The projection of the initial chromosomes and the searched Pareto solutions on the T_{max}-ω_{chara} plane for different combinations of N and g_{max} are illustrated in Fig. 6.11.

Similar to the results of MOPSO, the searched Pareto solutions with NSGA2 are concentrated to a narrow and limited region. A further increase of the chromosomes N or maximal iterations g_{max} cost obviously more computational time, but the improvement of the results is not significant any more.

Considering the inevitable randomness of both methods, the searched Pareto solutions from MOPSO and NSGA2 for N equal to 200 and and g_{max} equal to 1000 are combined together to generate a more widespread set of the PMTFM-CP designs for a better approximation of the real Pareto front, as shown in Fig. 6.11. It can be concluded that the results are satisfactory and acceptable in terms of the predefined criteria.

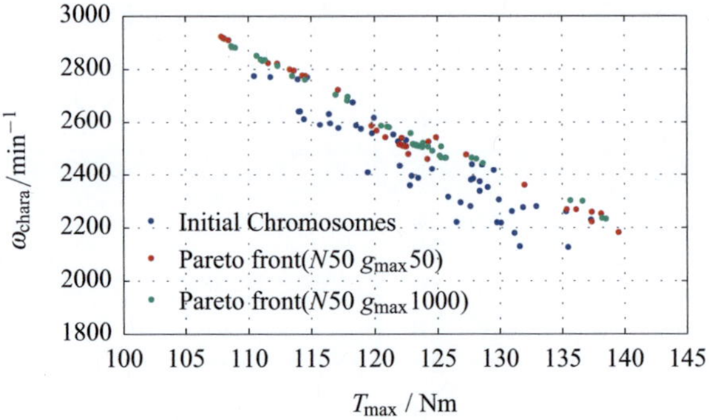

(a) 50 initial chromosomes

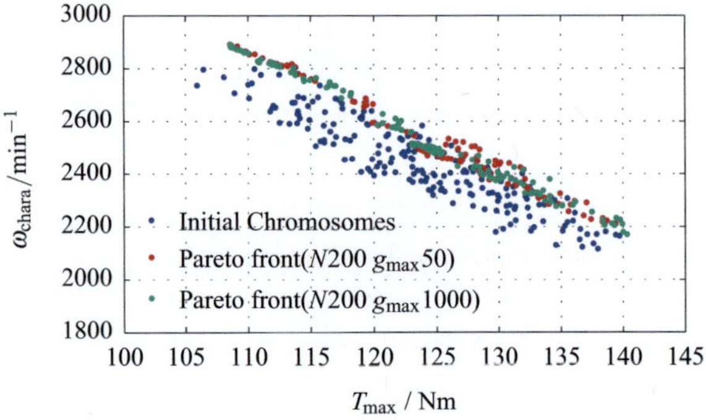

(b) 200 initial chromosomes

Figure 6.11. Initial chromosomes and Pareto front for different N and g_{max}

6.5 Optimal design selection based on fuzzy logic

Purpose of this section is to find out the optimal PMTFM-CP design with the best comprehensive performance from all the Pareto solutions in Fig. 6.12.

However, the performance of different PMTFM-CP designs cannot be directly compared as there are five objectives.

As a result, an evaluation method based on FL is utilized in this work, which can determine a crisp evaluating value for each Pareto solution in consideration of all five objectives. The idea of FL comes from the simulation of the human thinking, which consists of fuzzification, the fuzzy inference and the defuzzification.

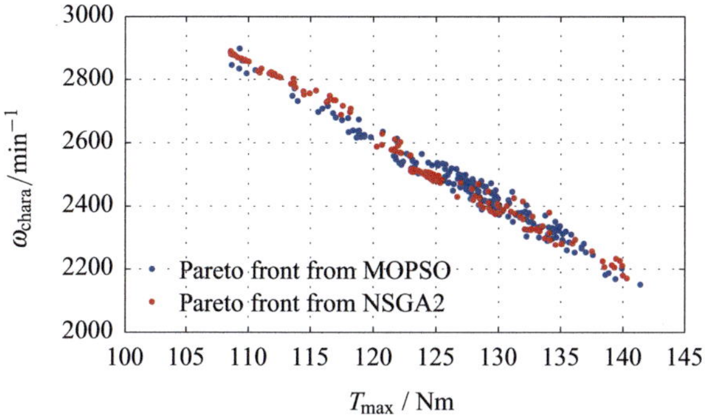

Figure 6.12. Combined Pareto solutions from MOPSO and NSGA2

The flow chart of FL is illustrated in Fig. 6.13.

- Firstly, the value range of each objective function f_i is divided into a few regions. Each region is associated with a linguistic label and a member ship function, whose value is named as membership degree and equals a value between 0 and 1. All the membership functions and the corresponding linguistic labels form the fuzzy sets. On this basis, the crisp value of each objective is transformed into the corresponding membership degrees with corresponding linguistic labels after the fuzzification.
- Similar to the fuzzy sets of the fuzzification, the rule base should also be predefined and is the fundament of the inference. It consists of many IF-THEN rules to associate the linguistic labels of the objectives and linguistic labels of the performance. With the decision making unit, the membership degree of performance linguistic labels are determined from that of the objective linguistic labels considering the specified logical operations. Afterwards, the values of all performance linguistic labels are aggregated. The above process is defined as fuzzy inference.

- The last step is defuzzification, in which a crisp value can be calculated from the above aggregation for each Pareto solution with the help of the centre of gravity method.

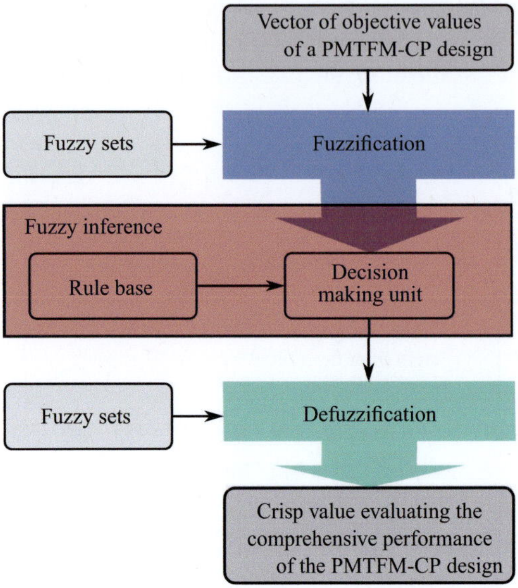

Figure 6.13. Flow chart of fuzzy logic

The above process has been carried out for each PMTFM-CP design to obtain a crisp value. By this means, the Pareto solutions can be compared with each other. The Pareto solution with the highest value is considered as the optimal PMTFM-CP design with the best comprehensive performance.

In this work, the selection based on FL is carried out with the help of the FL designer toolbox in MATLAB. The value range of each objective is divided into five regions, which correspond to five membership functions with different linguistic labels.

For instance, five triangular functions are defined as the membership functions of the objective $T_{\max}$ and the corresponding linguistic labels are "lower", "low", "normal", "high" and "higher" respectively, as illustrated in Fig. 6.14. In this case, the torque T_1 are relevant to two member ship functions with the linguistic labels "low" and "normal", whose membership degrees are equal to 0.4 and 0.6 respectively.

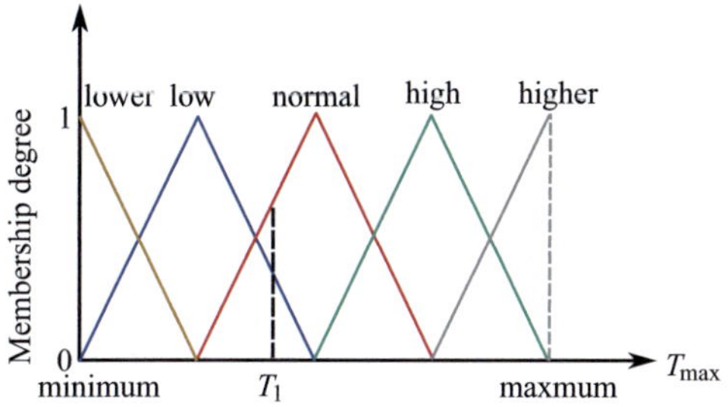

Figure 6.14. Membership functions to evaluate the objective T_{max}

Besides, totally 21 membership functions are defined to evaluate the performance of PMTFM-CP, whose linguistic labels are "level 0", "level 1", ... and "level 20" respectively, as illustrated in Fig. 6.15.

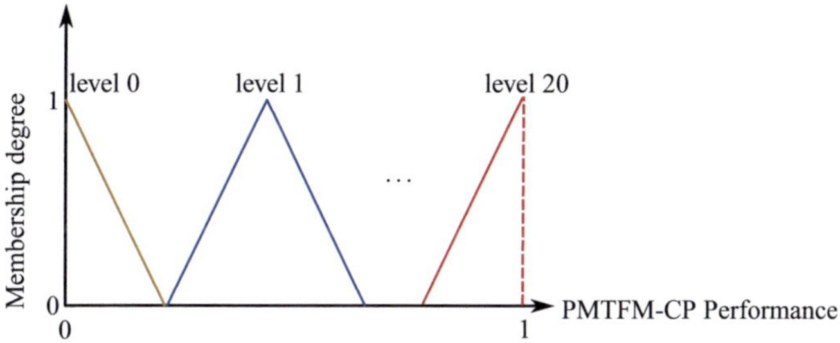

Figure 6.15. Membership functions to evaluate the PMTFM-CP

Subsequently, the Rule base is defined, which consists of 3125 IF-THEN rules considering all possible combinations of the 25 membership functions of the 5 objectives. For example, an IF-THEN rule is defined as follows.

IF T_{max} is "low", P_{cu} is "high", ω_{chara} is "normal", η_{chara} is "normal" and α_T is "normal", **THEN** the performance is "level 10".

The membership degree of the performance linguistic label "level 10" is calculated as the minimal value of the membership degrees of the five objective membership functions in the condition of the IF-THEN rule.

Generally, the conditions of 32 IF-THEN rules are fulfilled for each PMTFM-CP design. The values of all 21 performance linguistic labels are combined together based on the maximal principal in this work. Finally, a crisp value can be calculated for each PMTFM-CP design through the aggregation of all relevant membership functions based on the centre of gravity method.

The PMTFM-CP design with the highest crisp value can offer the best comprehensive performance, whose design parameters are listed in Table 6.8.

Parameter	**A** W_{basic}	**B** L_{tooth}	**C** W_{tooth}	**D** W_{claw}	**E** L_{claw}
Result	1	9.41 mm	0	0	0.859
Parameter	**F** L_{short}	**G** R_{short}	**H** R_{RYR}	**I** L_{PMR}	**J** p
Result	0.6	0.767	0.8	0.866	20

Table 6.8. Parameters of the final design of PMTFM-CP

The objective values of final PMTFM-CP design are listed in Table 6.9, which are validated with 3D FEM. It can be noted the FEM results match well with that of the mathematical model. For the final design, a high characteristic speed can be achieved at the cost of a relative low maximum torque.

Objective	T_{max} [Nm]	P_{cu} [W]	ω_{chara} [min^{-1}]	η_{chara} [%]	α_T [%]
Model	124.93	170.39	2523	95.85	9.75
FEM	125.58	171.14	2478	95.24	9.57

Table 6.9. Objective values of the final design of PMTFM-CP

7

Comparison of PMAFM-ISA, PMTFM-CP and PMRFM

In this chapter, the newly designed PMTFM-CP and the PMAFM-ISA are compared with the current design of PMRFM of Daimler AG. should be compared with each other. It is noted that the pole pair number of PMTFM-CP is determined as 20 to achieve a better compromise of the 5 objectives. In order to exhibit the high torque density of PMTFM, a PMTFM-CP with the number of turns of coil per phase equal to 6 and 30 pole pairs is also taken into account in the comparison, whose parameters are listed in Table 7.1.

Parameter	**A** W_{basic}	**B** L_{tooth}	**C** W_{tooth}	**D** W_{claw}	**E** L_{claw}
Result	1	8 mm	0.77	0	0.6
Parameter	**F** L_{short}	**G** R_{short}	**H** R_{RYR}	**I** L_{PMR}	**J** p
Result	0.8	1	1	1	30

Table 7.1. Parameters of the final design of PMTFM-CP

In addition, the final electromagnetic design of PMAFM-ISA with and without PM segmentation are also considered. In summary, the following topologies listed in Table 7.2 are compared in this section.

Topology	Description
PMAFM-ISA1	Final electromagnetic design of PMAFM-ISA
PMAFM-ISA2	The same electromagnetic design with PM segmentation
PMTFM-CP1	Optimized PMTFM-CP design with $p = 20$ (Table 6.8)
PMTFM-CP2	PMTFM-CP design with $p = 30$ (Table 7.1)
PMRFM	Current PMRFM design described in section 2.1

Table 7.2. Comparison of PM excited topologies

The PMAFM-ISA1 corresponds to the final electromagnetic design. In comparison, the PMAFM-ISA2 is the same design as PMAFM-ISA1, but the PM is segmented in 9 parts to reduce the eddy current. The PMTFM-CP1 is the optimized electromagnetic design based on the method described in 6, while the PMTFM-CP2 is a relatively arbitrary design whose parameters are listed in Table 7.1.

The 3D FEM models of PMAFM-ISA and PMTFM-CP are illustrated in Fig. 7.1, where the DLCW in PMAFM-ISA is faded out for a better overview. It should be noted that all the electrical machines are analysed with transient 3D FEM to consider the time-varying effects and to obtain more accurate results.

7.1 Cogging torque

One of the main disadvantages of the permanent magnet excited synchronous machine is the cogging torque, which is caused by the interaction between the PMs and the stator teeth. It can raise the additional vibrations, acoustic noise and complicated starting conditions. Thus, the cogging torque of the three topologies is compared firstly, as illustrated in Fig. 7.2.

Although the PM is skewed, the PMAFM-ISA shows the highest torque ripple due to the strong interaction between PM and stator. In addition, the cogging torque isn't obviously influenced by the segmentation of PM. Compared with PMTFM-CP1, the PMTFM-CP2 with higher pole pair number demonstrates higher cogging torque.

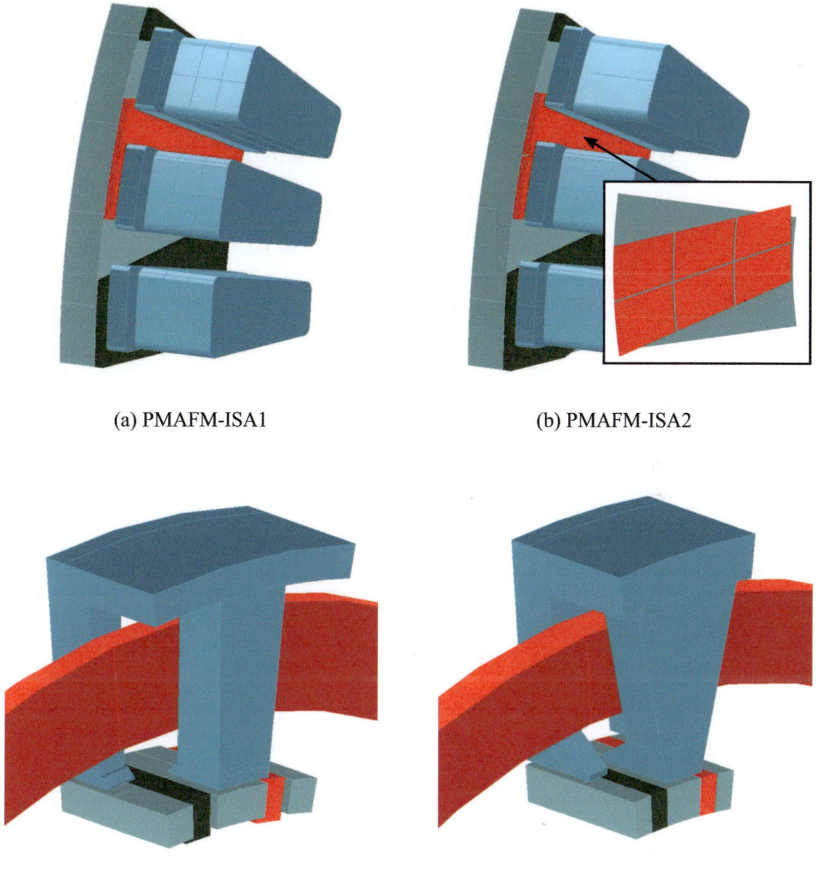

(a) PMAFM-ISA1

(b) PMAFM-ISA2

(c) PMTFM-CP1

(d) PMTFM-CP2

Figure 7.1. FEM models of different topologies for comparison

It is noted that the electrical machine with embedded PM, namely PMTFM-CP and PMRFM, has lower cogging torque than PMAFM-ISA with surface mounted PM, because the interaction between the PM flux and the stator teeth is effectively limited. The cogging torque of PMRFM is further reduced by means of the relative rotation of the 7 rotor segments, as illustrated in Fig. 2.2.

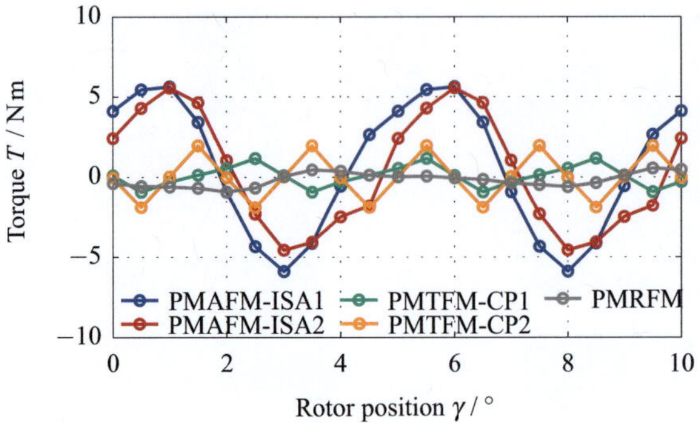

Figure 7.2. Comparison of cogging torque of different topologies

7.2 No-load losses

In this section, the losses of the above five electrical machines at no-load condition are compared, which consist of only the iron losses in SMC and the eddy current losses in PMs. The eddy current losses of PM and the total losses of all topologies at no-load are illustrated in Fig. 7.3.

Compared with the PMTFM-CP and PMRFM, the PMAFM-ISA demonstrates higher losses and the most important reason is the high eddy current losses generated in PMs. Through the comparison of the two topologies of PMAFM-ISA, it can be noted that the segmentation of PM is of great importance for the electrical machine with surface mounted PM.

Compared with the topology with surface mounted PM, the eddy current losses can be effectively restricted in the PMTFM-CP and PMRFM, where the PM are embedded in the rotor. Thus, a minor variation of the magnetic field in PM is achieved. Furthermore, the PMTFM-CP2 demonstrates higher losses than the PMTFM-CP1 due to higher pole pair number and the PMRFM shows the lowest losses. At last, the PMRFM has lower total losses than PMTFM-CP due to the lower hysteresis losses of electrical steel.

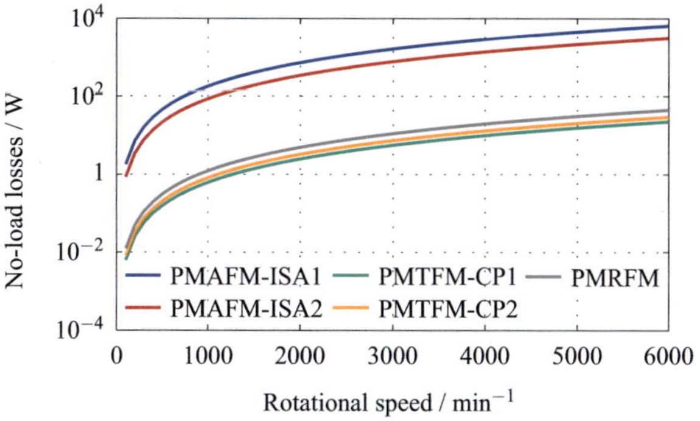

(a) Eddy current losses of PM

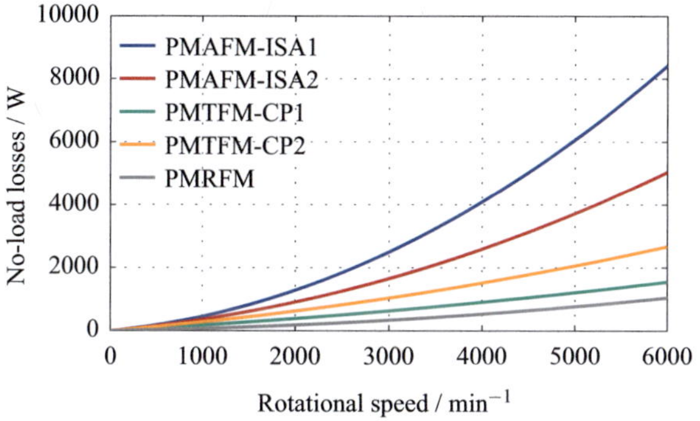

(b) Total losses

Figure 7.3. Comparison of different topologies at no-load

7.3 Average torque for constant current density

The average torque of an electrical machine at different electrical load is an important characteristic to evaluate its performance. For the case that there is only current component along the q- axis in the electrical machine, the average torque of different topologies for different J_q are compared, as illustrated in Fig. 7.4. Hereby, the filling factor of all topologies is assumed to be 0.7 for a better comparison.

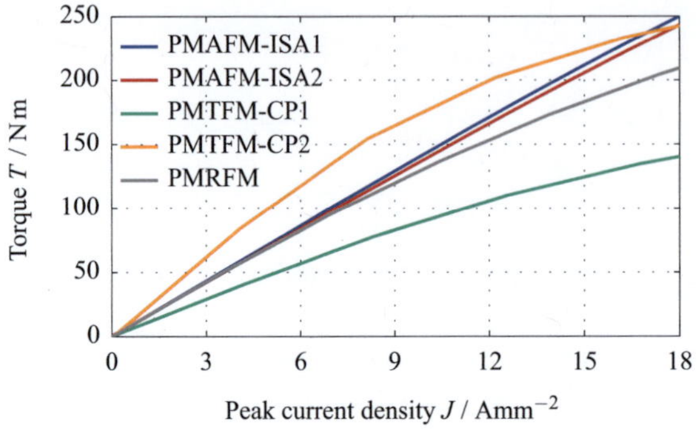

Figure 7.4. Comparison of torque of different topologies for the same J

Based on the comparison of two topologies of PMTFM-CP, it is apparent that the pole pair number has dominant influence on the average torque. The PMTFM-CP2 with 30 pole pairs demonstrates the highest torque when the electrical load is low, which becomes lower than PMAFM-ISA for the highest current density due to the high saturation, while the PMTFM-CP1 with 20 pole pairs offers the lowest torque compared with other topologies.

The topologies of PMAFM-ISA can offer higher average torque than the other topologies due to the large air gap area, which is not extremely saturated even for the highest current density. Compared with the PMAFM-ISA1, the output torque of PMAFM-ISA2 reduces slightly due to the segmentation of PM.

It can be seen that the output torque of PMRFM is lower than that of the PMAFM-ISA, which however can further increase for the same J by distributing the total current in d- and q-axis to utilize the reluctance torque.

7.4 Limiting characteristic curve

As one of the most important properties of electrical machine, the limiting characteristic curves of the five topologies are compared, which describes the peak output torque of electrical machine at different speed. For a better comparison, all the topologies are fed with the same inverter with parameters listed in Table 1.2 and the characteristic curves are illustrated in Fig. 7.5.

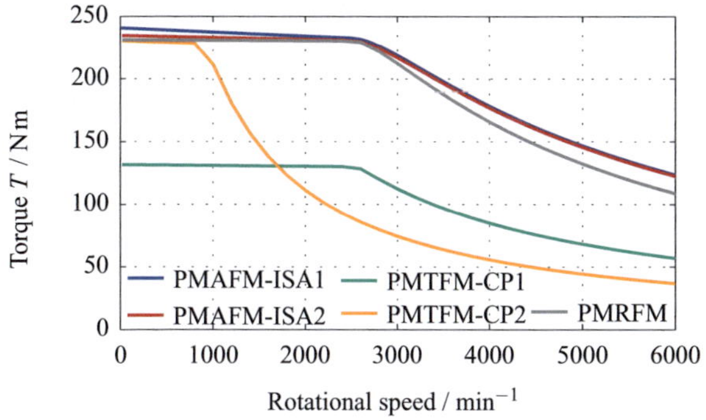

Figure 7.5. Comparison of limiting characteristic curve of all topologies

Obviously, the limiting characteristic curves of the two topologies of PMAFM are comparable, which are higher than that of PMRFM especially in the field weakening area. Although the segmentation of PM has no obvious influence on the characteristic curve, it can reduce the induced eddy currents losses in the PM as illustrated in Fig. 7.3, which is of great importance for the improvement of the continuous output power and the mechanical firmness of PM.

The PMTFM-CP2 can offer almost the same maximum torque as PMRFM, while its characteristic speed ω_{chara} becomes much smaller. In comparison, the characteristic speed ω_{chara} of PMTFM-CP1 is comparable with PMRFM, while the maximum torque is only about half of that of the PMRFM. In summary, the output power of PMTFM-CP is extremely limited due to the low power factor.

7.5 Efficiency map

First of all, the efficiency maps of PMAFM-ISA are illustrated in the Fig. 7.6. As the most informative method to evaluate an electrical machine, the efficiency map describes the optimal efficiency for all the operating points enclosed by the limiting characteristic curve. It is apparent that segmentation of PM leads to higher efficiency of PMAFM-ISA. For instance, the peak efficiency increases from 94.67 % to 96.04 % by this means.

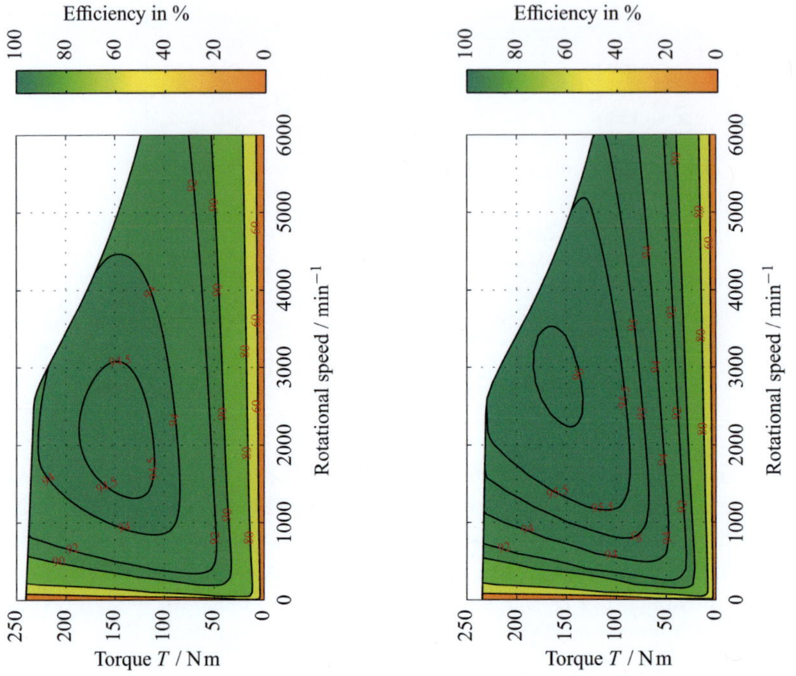

(a) PMAFM-ISA1

(b) PMAFM-ISA2

Figure 7.6. Comparison of efficiency maps of PMAFM-ISA

In comparison, the efficiency maps of PMTFM-CP are illustrated in Fig. 7.7. Both topologies can achieve higher efficiency than the PMAFM-ISA in large area, while peak efficiency of PMTFM-CP1 and PMTFM-CP2 are 97.43 % and 97.31 % respectively. Additionally, the PMTFM-CP1 demonstrates higher efficiency due to the lower losses owing to the lower pole pair number.

Finally, the efficiency map of the current PMRFM design is illustrated in Fig. 7.8. It can be noted that the peak efficiency equal to 96.5 % is lower than PMTFM, but a bit higher than the PMAFM-ISA with segmented PM.

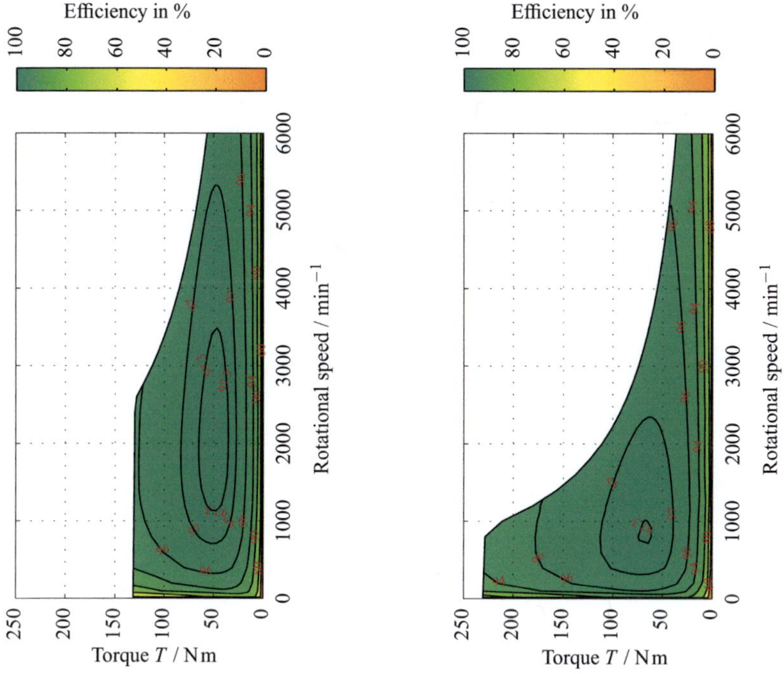

(a) PMTFM-CP1

(b) PMTFM-CP2

Figure 7.7. Comparison of efficiency maps of PMTFM-CP

However, the area in the efficiency map of PMRFM with efficiency higher than 90 % is equal to 72.24 %, while it equals 78.87 % and 72.98 % for the PMAFM-ISA with and without segmentation.

Based on the above analysis, the following conclusions can be drawn:

- The PM eddy current losses of PMTFM-CP and PMRFM are obviously lower than PMAFM-ISA as the PM are embedded in the rotor instead of being mounted on the surface of the rotor yoke.

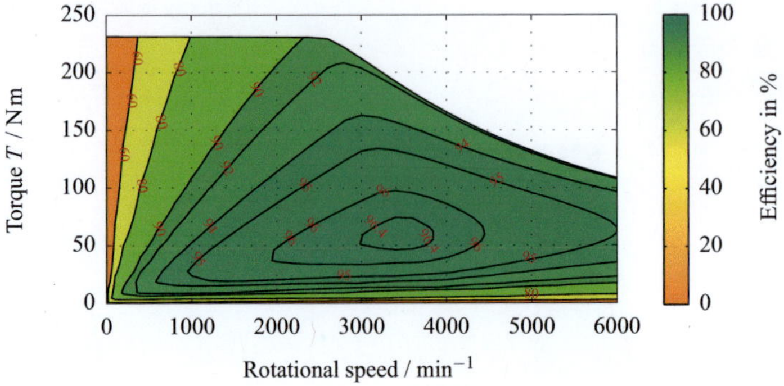

Figure 7.8. Efficiency map of PMRFM

- Compared with PMRFM, the PMAFM-ISA can offer higher peak output torque, especially at the field weakening area. However, the torque ripple of PMAFM-ISA is much higher than the PMTFM-CP and PMRFM.
- The number of pole pairs has a dominant influence on the output torque for PMTFM. For the same current density, the PMTFM with higher pole pair number has higher torque than PMAFM-ISA and PMRFM. On the other hand, the output power of PMTFM is much less than PMAFM-ISA and PMRFM for a specified inverter due to its low power factor.
- Relatively high efficiency can be achieved for all topologies due to the application of rare earth PM and windings with high filling factor. Through segmentation of PM, the efficiency of PMAFM-ISA can be effectively improved.

8

Mechanical Construction and Analysis

Through the comparison of the topologies described in chapter 7, it is apparent that the PMAFM-ISA is a suitable topology for the application of SMC. By this means, the complex stator teeth can be relatively easily manufactured to conduct the magnetic flux three-dimensionally.

In the chapter 5, the geometric parameters of all electromagnetic relevant components in the PMAFM-ISA have been determined. Subsequently, a prototype according to the PMAFM-ISA1 design should be constructed and manufactured to validate the FEM results [O10, S12]. It should be noted that the PMs aren't segmented in the first prototype to achieve better mechanical firmness.

The mechanical construction of PMAFM-ISA is very challenging because of mainly three reasons. First of all, the structure of PMAFM-ISA is very complex and the stator consists of many independent small segments, which should be fixed firmly during the rotation. Subsequently, the rotor yoke and the stator segments are made from SMC, which have inferior mechanical properties than conventional electrical steel. Considering the high costs, the SMC components aren't directly pressed, but milled from standard cylinder. In this case, the mechanical properties of SMC can get worse depending on the machining method. At last, the demountable ability is required for the first prototype to analyse the failure reason and to replace the broken components.

The chapter is organized as follows. First of all, the construction and analysis of the stator and rotor are described respectively. Subsequently, the modal and thermal analyses of the whole machine are performed in order to validate the suggested construction. Considering the complex structure of the machine, the coupled thermal and fluid dynamic (CFD) analysis is adopted to calculate the airflow and the thermal field based on the previously determined losses through the electromagnetic analysis.

The final electromagnetic design of PMAFM-ISA is illustrated in Fig. 8.1, whose parameters are listed in Table 5.12. The axial force depending on the rotor positions and the axial distance between the stator and the rotors is the basis of the mechanical analysis, while the total losses is the prerequisite for the thermal analysis. Both of them can be determined through the electromagnetic analysis.

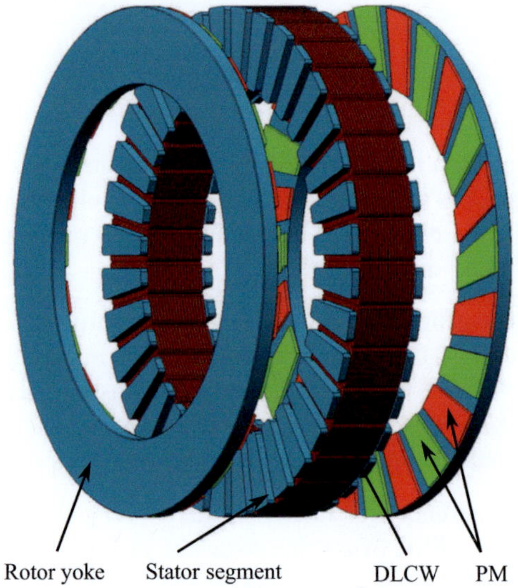

Figure 8.1. Final electromagnetic design of PMAFM-ISA with DLCW 12 × 2A1

8.1 Mechanical construction and analysis of stator

In this section, the mechanical construction and analysis of the stator are described [O10–O12, S13]. At first, the main challenges and the ultimate mechanical construction are described in details. Subsequently, the mechanical analysis of the stator based on FEM is presented.

8.1.1 Mechanical construction of stator

There are some special challenges and requirements for the mechanical design of the PMAFM-ISA prototype, which should be taken into account during the mechanical design of the stator:

- Despite the obvious improvement of SMC, the mechanical properties are still limited compared with electrical steel. For this reason, the stator segments and the rotor yoke cannot suffer large forces. A mechanical connection with help of the screws should not be utilized.
- Considering the short axial length of the air gap, the necessary space for the DLCW and the shaft, the small distance among the stator segments and the requirement of a compact stator, there is only very limited free space available.
- Because of the limited free space, the subsystems must be compact and robust, which must also work properly at high temperature.
- The machining has significant influence on the electromagnetic characteristics of the SMC components. For instance, drilling cannot be allowed.
- In order to reduce the assembly difficulty, the necessary manufacturing precision and machining difficulty, the number of the system components should be as few as possible. Simultaneously, the structure complexity and manufacturing costs should be limited in an acceptable level.
- A water-cooling system should be placed as closely as possible to the stator segments to reduce the thermal resistance, to dissipate the heat efficiently and to avoid the overheating of the stator.
- The resultant axial force acting on the stator is equal to zero in the ideal case, which can dramatically increase due to the asymmetric air gap length between the stator and the two rotors due to the inevitable manufacturing and assembly tolerances, as shown in Fig. 8.2.

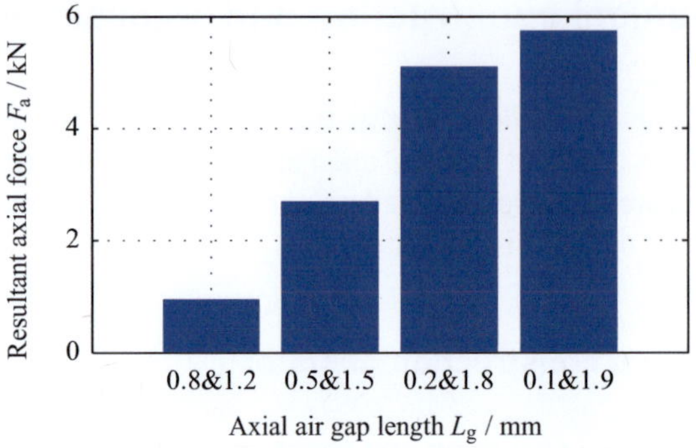

Figure 8.2. Resultant axial force on the stator for asymmetric air gap length

In consideration of the above requirements, the developed mechanical construction of the stator has been continuously improved and the ultimate version is illustrated in Fig. 8.3, where the stator segments belonging to the same phase have the same colour for a better clarity.

There are mainly four subsystems belonging to the stator. The first subsystem aims at accurate location and firm fastening of the stator segments during the operation. Subsequently, the water-cooling system is embedded in the stator to dissipate the generated heat resulted from diverse losses and to enable a higher electrical loading and continuous output power. Furthermore, the electrical system is necessary to distribute the current evenly in the 36 coils. The last subsystem is the housing, which is used to fix the stator and to protect the machine.

Hereby, all 36 stator segments with the concentrated winding are positioned and fixed with two annular plates. Each of them have 36 openings, which have the same shape as the stator shoe. By this means, the motion of the segments along the tangential and radial direction can be completely prevented. The deformation of the plates must be kept as small as possible during the operation because of the short axial length of the air gap, high elastic modulus and yield strength are required. In addition, a high temperature during the operation of the PMAFM-ISA can be achieved. Hence, a high allowable working temperature and a low thermal expansion coefficient are needed. Besides, the electrical resistivity should be high enough to limit the induced eddy currents in the plates. At last, a very low permeability is necessary so that the magnetic field is influenced.

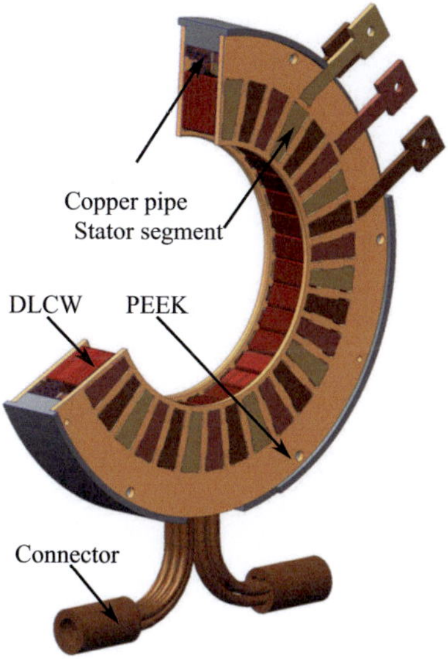

Figure 8.3. Cut view of the ultimate mechanical construction of stator

Considering the above requirements, some possible candidates belonging to polymer, technical ceramics and composites are selected and compared in terms of the allowable operating temperature and the electrical resistivity.

It can be seen that the material polyetheretherketon (PEEK) makes an excellent compromise between the electrical resistivity and the allowable working temperature. Taking the additional manufacturing difficulty and costs, the electromagnetic and mechanical properties into account, the PEEK is finally selected to produce the annular plates to guarantee the functionality of the first prototype, although the PEEK has high material price and manufacturing costs. The water cooling-system consists of five parallel connected copper pipes, which are connected with a temperature control system with two connectors. In order to dissipate the generated heat as soon as possible, the copper pipes are placed closely to the stator segments. In order to obtain a homogeneous temperature distribution, the five copper pipes are divided into two groups, where the cooling liquid flows in opposite direction.

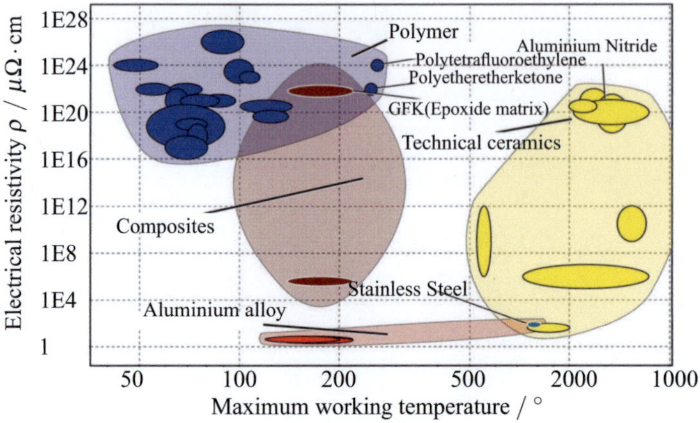

Figure 8.4. Comparison of different materials

The stator with housing and the electrical system are illustrated in Fig. 8.5, which consists of four copper rings and the corresponding bar connectors for the three phases in star connection and the neutral point.

With its help, the current from the inverter can be evenly distributed within the 36 coils. There are three phases in the PMAFM-ISA and each phase has six parallel paths. In general, it isn't necessary to export the neutral point for the electrical machine. However, with the accessible neutral point, the resistance and voltage of the three phases of the prototype can be easily measured, which is of great significance in examination of the functionality of the coils.

The housing is aimed to fix the bearings and to protect the internal components of the prototype, which is made from stainless steel with a similar thermal expansion coefficient to that of the bearing outer ring in order to guarantee a high precision of the bearing system. Considering its high density, 24 spindly holes are slotted on both sides of the housing to reduce the weight and to improve the cooling performance, as shown in Fig. 8.5.

In addition, the free space among the components of the stator is filled with epoxy resin with high thermal conductivity. By this means, the thermal resistance reduces significantly and all components are solidly assembled, so that the properties such as continuous power and robustness are improved. Therefore, the epoxy resin should possess high thermal conductivity, high mechanical strength and working temperature as well as low viscosity at room temperature.

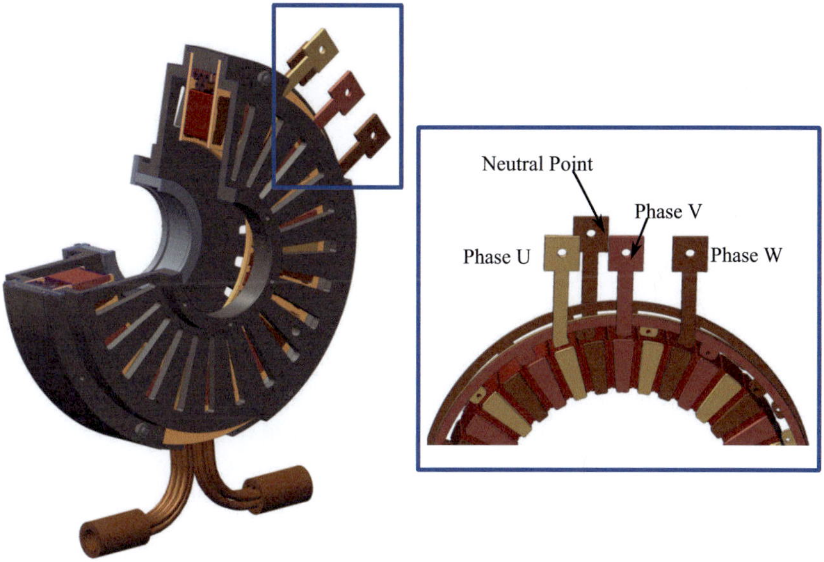

Figure 8.5. The stator with housing and electrical subsystem

8.1.2 Mechanical analysis of stator

Subsequently, the developed mechanical construction of the stator should be validated in terms of the stress and strain, which is performed based on the 3D FEM in order to achieve a high accuracy. Similar to the electromagnetic analysis, it is only necessary to analyse one pole pair model of PMAFM-ISA to reduce the simulation time.

The forces acting on the stator segments obtained from the Maxwell stress tensor are the basis for the calculation of the stress and deformation, which depend on the rotor position, the current in the windings and the axial length of the air gap. Thus, the mechanical analysis should be carried out for the maximum current and the most critical rotor position, where the resultant force on the stator segments reaches its maximum. In comparison, the radial force is small and negligible compared with the tangential F_t and axial force F_a, which are listed in Table 8.1.

Air Gap	**Segment** U_1		**Segment** V_1		**Segment** W_1	
[mm]	F_t[N]	F_a[N]	F_t[N]	F_a[N]	F_t[N]	F_a[N]
0.2	16.54	139.22	68.91	306.84	12.57	320.89
1	12.41	74.82	55.91	190.86	11.73	210.56
1.8	9.64	48.55	46.43	133.57	11.24	154.95

Table 8.1. Forces acting on stator segments depending on air gap length

It can be noted the forces are strongly depending on air gap length. For the asymmetrical air gaps, a large axial force can act on the stator. For instance, the equivalent stress and the total deformation of the stator segments are illustrated in Fig. 8.6 for the case that the air gap length equal to 0.2 mm and 1.8 mm.

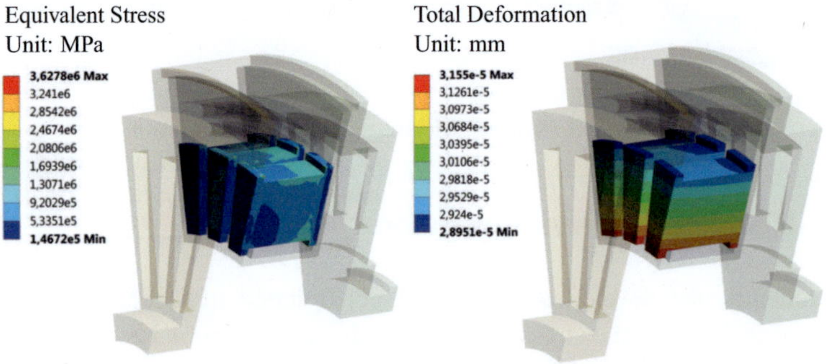

Figure 8.6. Equivalent stress and deformation of the stator segments

The tensile stress R_y, the ultimate stress R_m, the maximum equivalent stress σ_{max} and maximum total deformation δ_{max} of the crucial components of the stator for symmetrical and asymmetrical air gaps corresponding to the forces in Table 8.1 are listed in Table 8.2.

From the above analysis, it can be noted that both the stress and deformation increase significantly, when the air gaps become asymmetric. However, even for the worst case that the air gaps are equal to 0.2 mm and 1.8 mm respectively, the maximum stress is much lower than the ultimate stress of the brittle materials and the tensile stress of the ductile material, while the deformation is much lower than the air gap length. In other words, the developed construction of stator is valid from the perspective of mechanical analysis.

Component	R_y	R_m	Symmetric		Asymmetric	
			σ_{max}	δ_{max}	σ_{max}	δ_{max}
	[MPa]	[MPa]	[MPa]	[mm]	[MPa]	[mm]
Stator segments	–	35	1.46	1.28E – 4	3.63	3.16E – 2
Steel housing	440	640	0.179	1.02E – 4	29.99	2.83E – 2
PEEK plates	95	–	0.063	1.27E – 4	0.91	3.17E – 2
Epoxy resin	–	66	0.31	1.10E – 4	0.94	3.15E – 2

Table 8.2. Maximum stress and deformation of stator components for different air gaps

8.2 Mechanical construction and analysis of rotor

In this section, the mechanical construction and analysis of the rotor are explained. At first, the design of the bearing system and the selection of the bearing are explained in details. On this basis, the construction of the complete rotor is introduced. Subsequently, the adhesive performance test has been carried out in order to investigate the maximum stress of the adhesive used in the construction. At last, the determination of the fit and the mechanical analysis based on FEM are presented.

8.2.1 Design of bearing system

Similar to the stator, the biggest challenge for the construction of the rotor is the large axial force. For the symmetrical air gaps, there is no axial force acting on the bearings. However, as already emphasized, there will be high force for the asymmetric air gaps and in this case, the bearings must be able to sustain large axial force and should possess high axial stiffness and accuracy. In addition, the shaft distortion resulted from the thermal expansion should also be taken into account to avoid high thermal stress. At last, a large inner diameter of the bearings is preferred to improve the rigidity and deflection of the prototype.

Based on the above considerations, a fixed-floating bearing arrangement has been selected for the prototype. The fixed bearings should withstand the radial and axial force and the shaft should be fixed axially in both directions. Furthermore, the fixed bearings must have suitably high precision and rigidity and is capable of sustaining large axial force. on the other hand, the floating bearing should only provide radial support and not be fixed to enable the free displacement of the shaft to avoid thermal stress.

The bearing system is illustrated in Fig. 8.7. Considering the above requirements, two precise spindle bearings B71914-E-2RSD-T-P4S-DUL have been used as fixed bearings and the floating bearing is a conventional deep groove ball bearing 6014-2Z.

In order to avoid designing of the complex lubrication system and to reduce the system maintain costs, all the bearings are sealed on both sides and lubricated with grease for a whole lifetime.

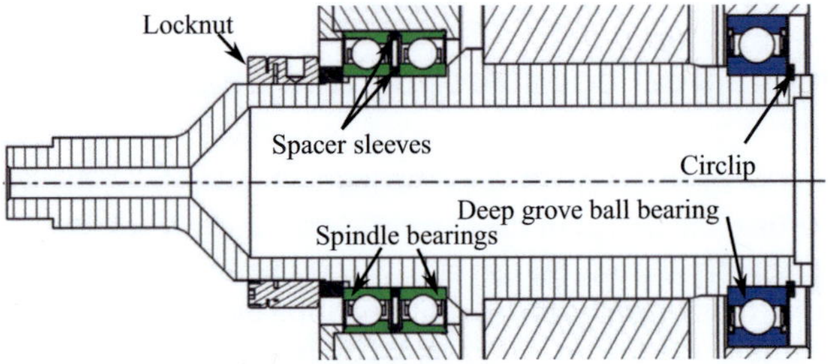

Figure 8.7. Bearing system of PMAFM-ISA

The rigidity of the bearing system can be obviously influenced by the preload, the arrangement and the contact angle of the spindle bearings. In general, a high preload leads to high rigidity and high lift-off force, which is defined as the external axial force when a bearing is relieved of preload. Nevertheless, the maximum allowable speed decreases significantly when the bearings are fitted with higher preload. Considering the axial force listed in Table 8.1, a preload equal to 890 N is performed on the two spindle bearings, which is realized with a locknut and two spacer sleeves with different thickness located between the two spindle.

The spindle bearings have large contact angle equal to 25°, to achieve a high axial rigidity. On the other hand, an outstanding compromise between the axial and radial load carrying capacity can be achieved. Furthermore, a high temperature difference between the inner and outer ring of the bearing becomes permissible.

Finally, the two spindle bearings are placed in an O-arrangement to raise the allowable rotational speed.

With the help of the online tool BEARINX® developed by the firm Schaeffler, some important characteristics such as fatigue life can be determined in consideration of many factors such as the preload acting on the spindle bearings, the elastic properties of the shaft and so on. The model of the shaft with the bearings are illustrated in Fig. 8.8.

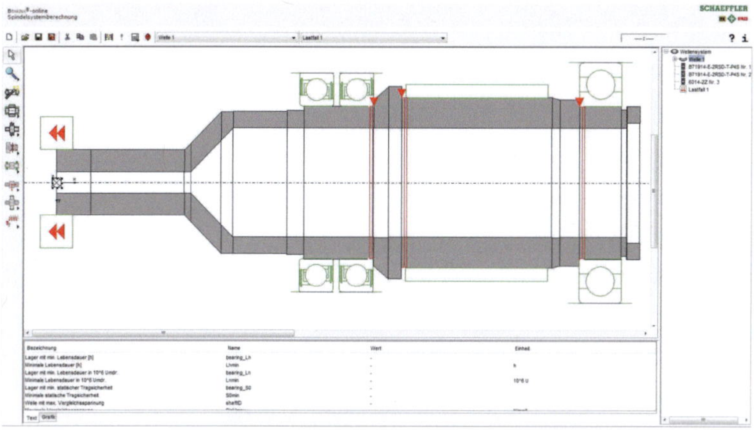

Figure 8.8. Online-tool BEARINX® for bearing calculation

The nominal reference rating life of the spindle bearing and the deep grove ball bearing depending on the external axial force resulted from the asymmetric air gaps is calculated for the maximum torque 230 Nm and rotational speed 7000 rpm and listed in Table 8.3.

Axial force	**Nominal reference rating life**	
	Spindle bearing	**Deep grove ball bearing**
[N]	[h]	[h]
1600	13929	6841
2000	8067	5988
3000	2718	4194
6000	384	1307

Table 8.3. Nominal reference rating life of bearings depending on the axial force

It can be noted that the nominal reference rating life reduces significantly due to the increasing axial force and the reduction of preload. Considering that the prototype needs to be measured in the test bench for less than 100 hours, it can be concluded that the suggested bearing system of the prototype is enough even for the worst case.

8.2.2 Construction of rotor

The most important challenges for the construction of the rotor are the limited mechanical properties of SMC, high axial force for the asymmetric air gaps and the requirements for simple manufacturing and assembly. In addition, it should be noted that the available size of the pressed standard cylinder is limited, so that the rotor yoke cannot be manufactured as a complete part. Instead, each rotor yoke is divided into 12 segments for the prototype.

Based on the above considerations, the construction of the rotor has been developed and continuously improved. The ultimate design is illustrated in Fig. 8.9.

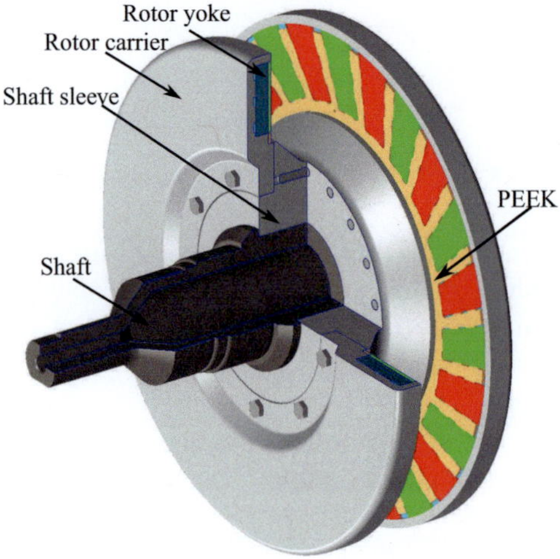

Figure 8.9. Cut view of the ultimate mechanical construction of rotor

The 12 segments of each rotor yoke are held by a rotor carrier made from aluminium alloy ENAW6082T6 with high mechanical strength. Both rotor carriers are bounded together with the shaft sleeve made from the same material through shrinking, which are additionally screwed together for a better reliability.

In order to guarantee the precision of the spindle bearings, the shaft is made from the steel 16MnCr5, whose thermal expansion coefficient is similar to that of the inner ring of the spindle bearing. The shaft sleeve is bounded together with the shaft through shrinking. In order to reduce the weight and the deflection of the rotor during the rotation caused by the asymmetric loads, a hollow shaft with large outer diameter is designed.

In addition to the electromagnetic force, the rotor yoke and PMs suffer from large centrifugal force, especially at high rotational speed. For this reason, the aluminium alloy with outstanding mechanical properties and low density is used to manufacture the rotor carriers and the shaft sleeve, whose structure has also been optimized to reduce the deformation.

The PM mounted on both rotor yokes should be exactly aligned with each other, which is realized with the help of the two annular plates made form PEEK. Another function of the plates is to support the PM in tangential direction during the rotation.

8.2.3 Adhesive performance test

As the PMs, the rotor yoke and the rotor carrier are bounded together with help of a technical adhesive, it is necessary to investigate the maximum allowable stress of the utilized adhesive at different temperature.

Because the property of the adhesive becomes worse at high temperature, the measurement should be carried out at different temperature. The designed devices and samples for the adhesive performance test are illustrated in Fig. 8.10.

The sample consists of a SMC plate and a PM plate, which are glued together with the used industrial adhesive. At first, the sample is placed in the container filled with silicone oil, which is heated to the desired temperature.

Subsequently, the measuring detector moves vertically and slowly down to apply large force on the SMC plate. The force is continuously measured with the detector until the two plates are separated from each other. For instance, the force measured with the detector for a test performed at 20° is illustrated in Fig. 8.11.

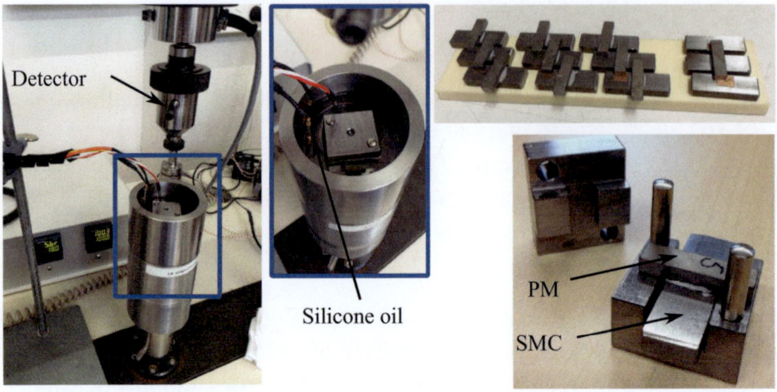

Figure 8.10. Device and samples for the adhesive performance test

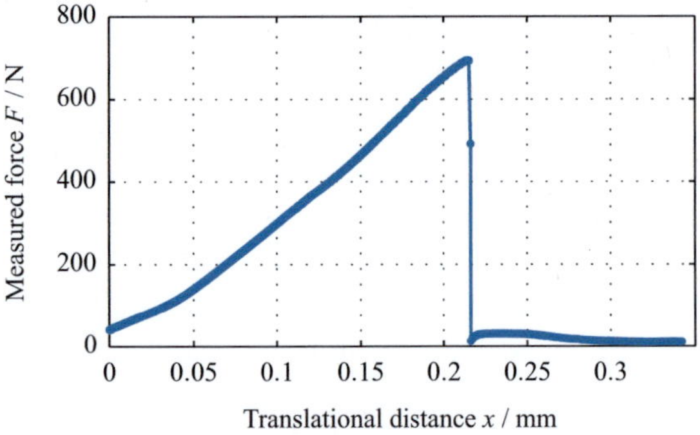

Figure 8.11. Measured force during the adhesive performance test

In order to eliminate the randomness of the test and to obtain a more reliable conclusion, at least four samples are measured for each temperature. The measured maximum stress at different temperatures are listed in Table 8.4. It can be noted that the maximum allowable stress of the adhesive reduces significantly with increasing temperature, which should not be exceeded during the operation of the prototype.

	Temperature [°C]				
	20	**100**	**120**	**140**	**180**
Maximum stress [MPa]	14.59	12.23	7.26	2.58	1.27

Table 8.4. The maximum allowable stress of the adhesive at different temperatures

8.2.4 Mechanical analysis of rotor

The rotor carriers, shaft sleeve and main shaft are bounded together through shrinking. In order to achieve high reliability during rotation, the tolerance and fit between the components should be determined with great attention.

It should be noted the most critical connection is the fit between the shaft sleeve made from aluminium alloy and the shaft made from steel, which has obviously lower thermal expansion coefficient. For this reason, a shrink fit with high fit excess has been designed and validated with the software KISSsoft, to obtain a firm connection between the two components even for the most serious case, where the maximum torque equals 230 Nm at rotational speed 7000 rpm under an axial external force 6000 N.

Components	**Fit**	**Bonding temperature** ΔT
shaft sleeve & shaft	x6/H7	$133°C$
shaft sleeve & rotor carrier	t6/H7	$91°C$

Table 8.5. Determined fits of the prototype with KISSsoft

Considering the high pressure resulted from the above shrink fit, it is necessary to analyse the stress of the main shaft, as illustrated in Fig. 8.12. It is noted that even for the worst case the maximum equivalent stress is much lower than the tensile yield stress of the steel 16MnCr5 equal to 440 MPa and the design of shaft is valid.

Subsequently, the form of the rotor carrier is optimized in terms of the deformation, weight and stress. Similar to the stator, the forces acting on the PM are depending on the current, the axial length and the rotor position as well, which can be determined with the electromagnetic analysis. For the rotor position with the largest resultant force acting on the PM, the results are listed in Table 8.6.

Figure 8.12. Equivalent stress of the main shaft

Air Gap	PM 1		PM2	
[mm]	F_t[N]	F_a[N]	F_t[N]	F_a[N]
0.2	68.25	329.5	29.77	437.44
1	54.56	234.44	25.49	241.81

Table 8.6. Forces acting on the PM for different air gap length

Based on the forces in Table 8.6, the mechanical analysis of the rotor has been carried out. For instance, the deformation of the rotor carrier and the stress of the PM are illustrated in Fig. 8.13, which correspond to the maximum torque 230 Nm at the characteristic rotational speed 2400 rpm for the asymmetrical air gap length 0.2 mm and 1.8 mm.

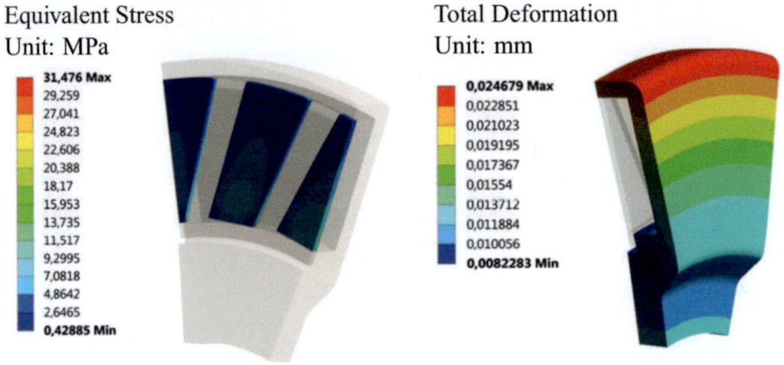

Figure 8.13. Equivalent stress and deformation of the rotor components

Similar to the stator, the tensile stress R_y, the ultimate stress R_m, the maximum equivalent stress σ_{max} and maximum total deformation δ_{max} of the crucial components of the rotor for symmetrical and asymmetrical air gaps (0.2 mm and 1.8 mm) are listed in the following Table 8.7. It can be seen that all components of the rotor are valid in terms of the deformation and the stress. In other words, the construction of the rotor is feasible and reliable.

Component	R_y	R_m	**Symmetric**		**Asymmetric**	
			σ_{max}	δ_{max}	σ_{max}	δ_{max}
	[MPa]	[MPa]	[MPa]	[mm]	[MPa]	[mm]
Rotor yoke	–	35	20.39	1.39E − 2	20.52	2.26E − 2
Rotor carrier	250	280	17.19	1.51E − 2	27.03	2.47E − 2
PEEK plate	95	–	0.79	1.24E − 2	1.21	2.02E − 2
PM	–	98	27.19	1.28E − 2	31.48	2.12E − 2

Table 8.7. Maximum stress and deformation of rotor components for different air gaps

In order to reduce the simulation time and the mesh error, the extremely thin adhesive with axial length of 0.1 mm is not directly modelled in the complete model of the whole rotor, but investigated separately in a simplified model including only a PM with the adhesive.

The stress of the adhesive corresponding to an axial force equal to 500 N at the speed 7000 rpm is illustrated in Fig. 8.14. Considering the maximal stress equal to 2.88 MPa and the measured results in Table 8.4, it can be concluded that the maximum temperature of the adhesive should not be higher than 120°C.

Figure 8.14. Equivalent stress of the adhesive

8.3 Final mechanical construction and modal analysis

The complete mechanical construction of the prototype can be achieved by combining the above constructed stator and rotor as well as some other important components such as resolver and wave spring washer. The cross section and exploded view of the final mechanical construction are illustrated in Fig. 8.15.

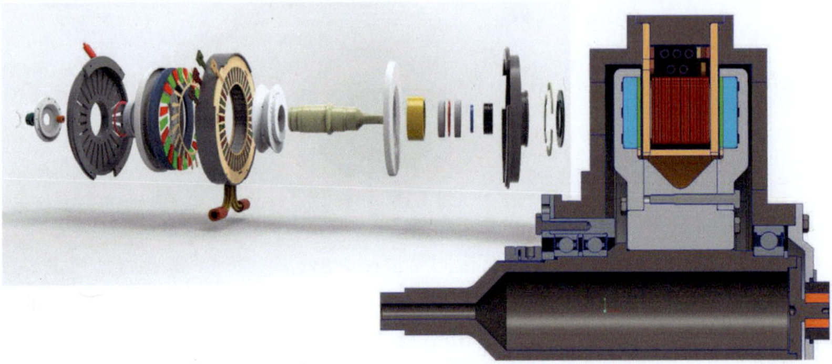

Figure 8.15. Cross section and exploded view of final construction

Based on the complete construction of the prototype, the model analysis has been carried out to investigate the free vibration behaviour of the dynamic characteristics. Because the complex structure of the bearings leads to unacceptable computation time, they cannot be exactly modelled, but modelled as solid cylinders. The equivalent elastic modulus of the cylinder is determined so that it has the same deformation as the real bearing under the same axial force.

The first four modes and the corresponding frequencies are illustrated in Fig. 8.16. The frequency of the first mode equal to 239.23Hz corresponds to a rotational speed of 10150 rpm, which is much higher than the required maximum speed and implies the resonance cannot occur.

8.4 Thermal analysis

As mentioned above, the properties of both adhesive and PM are strongly depending on the temperature.

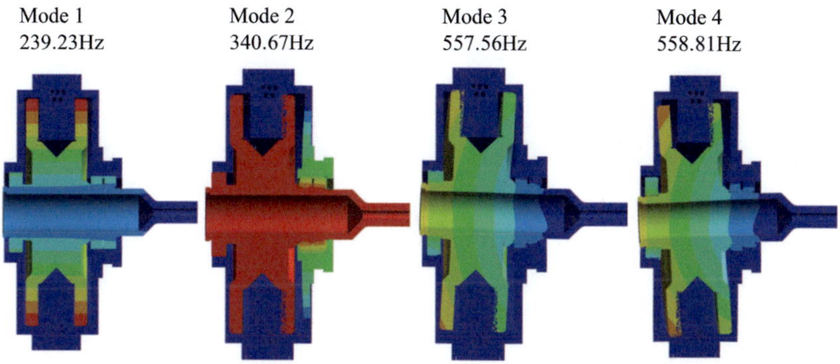

Figure 8.16. Vibration modes and frequencies of the PMAFM-ISA

At temperature higher than 120 °C, the adhesive can lose its functionality leading to the shedding of PM. Besides, the remanence and coercive field of the PM reduce with increasing temperature, which can even demagnetized at extremely high temperature for large field weakening current [S14]. As a result, the temperature is extremely critical for the functionality and performance of the prototype and should be carefully investigated [S15].

Due to the complex structure of the PMAFM-ISA, the coupled thermal and computational fluid dynamics (CFD) analysis is utilized to determine the airflow and the temperature of the electrical machine. As the basis of the thermal analysis, the total losses corresponding to the heat sources are calculated with the electromagnetic analysis.

Comparison with electromagnetic analysis, CFD analysis is much more time-consuming as complete model should be analysed to determine the air flow inside and around the machine. For this reason, an analytical model is necessary in order to calculate the temperature more efficiently with less calculation time.

In this section, the coupled thermal and CFD analysis of the prototype with the software Star-CCM+ is firstly presented. Subsequently, an analytical model based on the results of CFD analysis developed with the software Matlab/Simulink is presented, which can predict the temperature with reasonable accuracy for a short computation time.

8.4.1 Coupled thermal and CFD analysis based on finite volume method

The thermal exchange in an electrical machine consists of three mechanisms including conduction, radiation and convection. The radiation is usually neglected considering the relatively low temperature of electrical machine, while the convective heat transfer effect is a very complex process due to the dependence of the exchanged heat on air flow and the dependence of air density on the pressure and temperature. For this reason, the software Star-CCM+ based on finite volume method is utilized in this work, which is able to accurately solve the coupled thermal and fluid-dynamics analysis.

In order to improve the flexibility, a parameterized model has been developed with Star-CCM+, which enables a rapid adjustment of the geometry of the PMAFM-ISA. In addition, the complete analysis process is controlled by a set of macros created with Java, so that a high efficient simulation can be realized. The cross section of the meshed model is illustrated in Fig. 8.17.

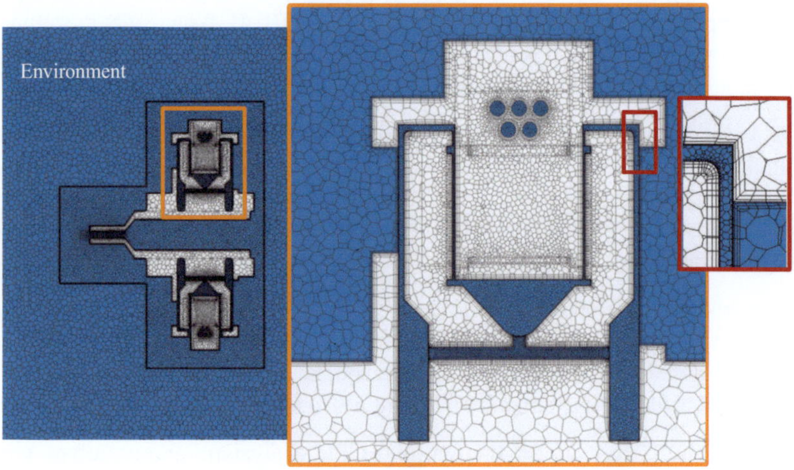

Figure 8.17. Cross section of the meshed model of CFD analysis

It can be seen that the components of the model have different mesh sizes to achieve a good compromises between high accuracy and calculation time. The air gaps have the smallest mesh size, where the air flow is highly turbulent, while the cells of the solid bodies have a relatively large size. In addition, the interfaces between the solid objects and the air are modelled with three prism layers to calculate the convection more accurately. At last, the model of the surrounding air around the prototype should be large enough to consider the exchanged heat and air flow in the ambient.

Because the coupled thermal and CFD analysis of the complete PMAFM-ISA is very time-consuming, only the stationary analysis is carried out in this thesis. In order to reduce the calculation time, the fluid in the cooling system of the stator is modelled as laminar flow. For instance, the axial and radial components of the air flow for PMAFM-ISA at the rotational speed $2600\,\mathrm{min}^{-1}$ are illustrated in Fig. 8.18. In comparison, the tangential component is much larger and therefore hidden for a better clarity.

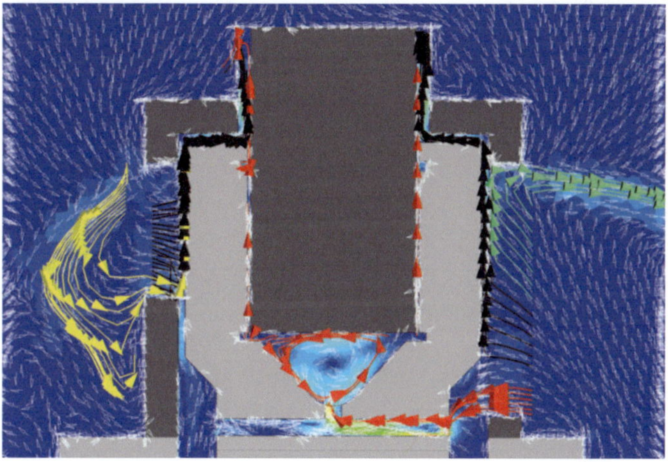

Figure 8.18. Axial and radial components of the air flow at $2600\,\mathrm{min}^{-1}$

Totally, three possible paths for the influent air flow can be well seen. Part of the air marked with red colour flows through the spindly holes of the right housing, the holes in the rotor carrier, the air gaps and finally the opening between stator and housing. Because the slotted holes in the left housing are smaller and located more external, the air can hardly flow into the prototype through the holes on the left housing. Overall, the air flow through the air gaps is very limited.

The second part of the air marked with black colour flows through the slots of housing into the PMAFM-ISA, which afterwards flows along the back side of the rotor carrier to the environment through the opening of housing. The last part of the air flowing into the prototype flows directly to the environment through the slotted holes.

For instance, the temperature field for the torque equal to 115 N m at rotational speed 2600 min^{-1} is illustrated in Fig. 8.19. In this case, the highest allowable temperature 120 °C of the PM is reached, which means the continuous power is nearly 50 % of the peak power at 2600 min^{-1}. Besides, it can be seen that the internal parts of the stator have the highest temperature due to the long distance to the water-cooling system and the poor air cooling.

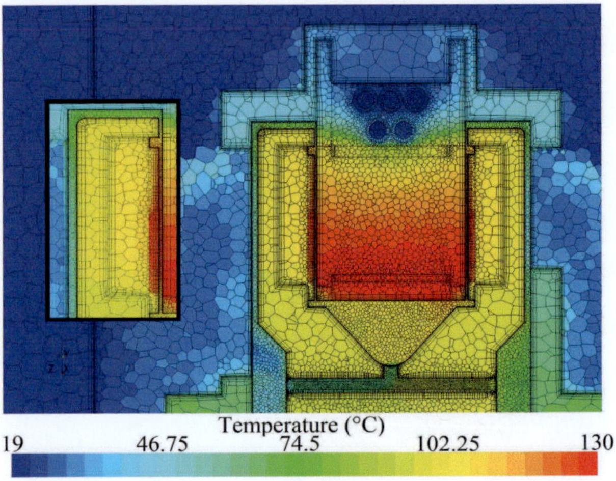

Figure 8.19. Temperature of PMAFM-ISA at 2600 min^{-1} and 115 N m

The accurate coupled thermal and CFD analysis is so time consuming that the method cannot be used to calculate the thermal field of PMAFM-ISA for all operating points. As a preparation of the following analytical thermal analysis, the air flows are measured at some locations of the model for different rotational speed to consider the transferred heat, as illustrated in Fig. 8.20. The determined mass flow of air for different rotational speeds are listed in Table 8.8.

It is apparent that only a small amount of air flows through the air gaps. In addition, it should be noted that the air through measurement locations 5 and 7 flows to the environment again through the housing opening, which cannot be observed in the current cross section. The same conclusion is also valid for the air flowing through locations 6 and 8.

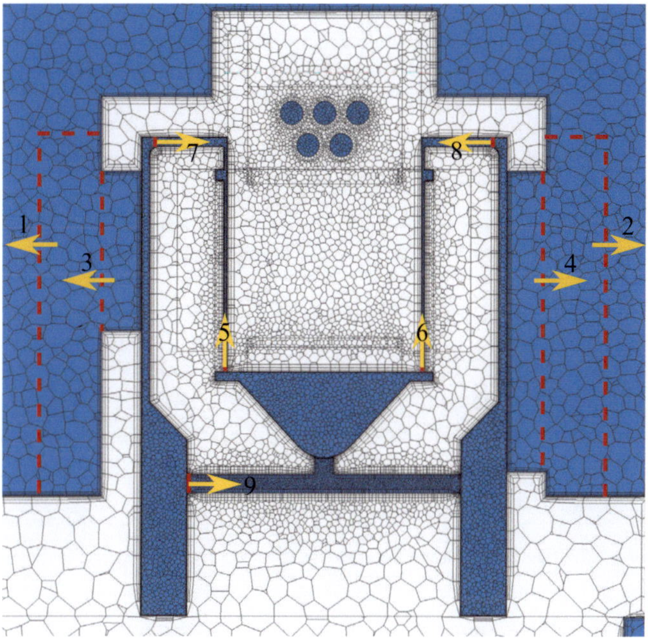

Figure 8.20. Measurement location of the air flow in CFD analysis

Speed	**Mass flow** [g/s]								
[rpm]	**1**	**2**	**3**	**4**	**5**	**6**	**7**	**8**	**9**
1000	4.75	7.62	3.99	5.19	0.25	0.21	1	1.19	0.003
2500	11.94	17.87	8.59	11.69	0.48	0.66	2.54	2.88	0.031
4000	19.58	27.45	13.5	18.05	0.84	1.24	3.89	4.29	0.057

Table 8.8. The air mass flow at different rotational speed

8.4.2 Thermal analysis with analytical model

Although the coupled thermal and fluid dynamic analysis offers highly accurate results, it is very time-consuming. Therefore, an analytical model has been developed in Matlab/Simulink to determine the temperature of the important components of PMAFM-ISA for all operating points, which is a thermal network based on the utilization of analogy to electrical circuit presented in Table 8.9.

Electrical		**Thermal**	
Voltage	U	Temperature	ΔT
Current	I	Heat flux	$\dot{Q}$
current density	J	Heat flux density	$\dot{q}$
Electrical Capacity	C	Thermal capacity	C_{th}
Electrical Resistance	R	Thermal resistance	R_{th}
Electrical conductivity	σ	Thermal conductivity	λ

Table 8.9. Analogy of the electrical and thermal quantities

The Fourier's law describes the thermal conduction within a solid body or between contiguous bodies resulted by the temperature gradient, where the $\dot{q}$ is the local heat flux density.

$$\dot{q} = -\lambda \cdot \nabla T = -\lambda \cdot \left(\frac{\mathrm{d}T}{\mathrm{d}x} + \frac{\mathrm{d}T}{\mathrm{d}y} + \frac{\mathrm{d}T}{\mathrm{d}z}\right) \tag{8.1}$$

For the simple application of a cuboidal element with constant λ and one-dimensional heat flux, the thermal resistance can be calculated using 8.2, where the d and A are the length and area of the sample.

$$R_{th} = \frac{d}{A} \tag{8.2}$$

In similar to the Ohm's law, the thermal resistance can be calculated with the Newton's law of cooling.

$$R_{th} = \frac{\Delta T}{\dot{Q}} \tag{8.3}$$

Based on above analysis, the thermal network of the PMAFM-ISA is developed based on the 3D thermal modelling of three basic solid elements, namely the cuboid, the hollow cylinder and the hollow cylinder segment.

For the thermal modelling of the basic elements, it is assumed that the heat flux along different directions are independent. In addition, the thermal mass and the average temperature of the element are the same for all directions.

Cuboid

The equivalent thermal network of a cuboidal element is illustrated in Fig. 8.21, which consists of three T-shape equivalent circuits for each direction. The thermal resistances can be calculated as:

$$R_{1x,y,z} = R_{2x,y,z} = \frac{l_{x,y,z}}{2 \cdot \lambda_{x,y,z} \cdot A_{x,y,z}} \tag{8.4}$$

$$R_{3x,y,z} = -\frac{l_{x,y,z}}{6 \cdot \lambda_{x,y,z} \cdot A_{x,y,z}} \tag{8.5}$$

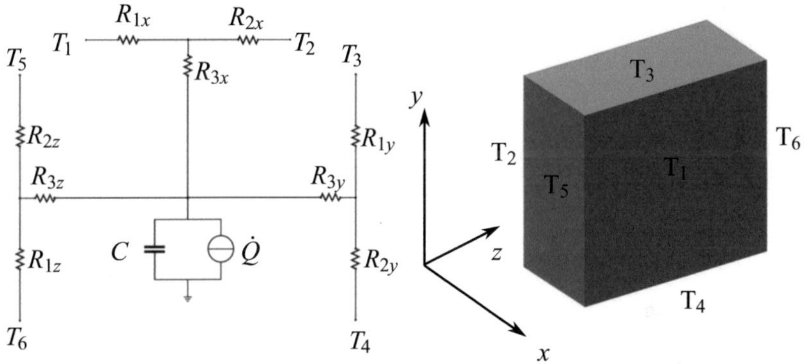

Figure 8.21. Equivalent thermal network of a cuboid

Hollow cylinder

The equivalent thermal network of a hollow cylinder comprises two T-shape equivalent circuits for the axial and radial direction respectively, as illustrated in Fig. 8.22.

The thermal resistances relevant to the axial direction can be calculated using the following equations:

$$R_{1z} = \frac{l_z}{2 \cdot \lambda_z \cdot A_z} \tag{8.6}$$

$$R_{2z} = \frac{l_z}{2 \cdot \lambda_z \cdot A_z} \tag{8.7}$$

$$R_{3z} = \frac{l_z}{6 \cdot \lambda_z \cdot A_z} \tag{8.8}$$

In comparison, the radial resistances towards outer and inner surface of the cylinder are not identical because of different equivalent area.

$$R_{1r} = \frac{1}{4 \cdot \pi \cdot \lambda_r \cdot l_z}\left[1 - \frac{2 \cdot r_i^2 \cdot ln(\frac{r_a}{r_i})}{r_a^2 - r_i^2}\right] \tag{8.9}$$

$$R_{2r} = \frac{1}{4 \cdot \pi \cdot \lambda_r \cdot l_z}\left[\frac{2 \cdot r_a^2 \cdot ln(\frac{r_a}{r_i})}{r_a^2 - r_i^2} - 1\right] \tag{8.10}$$

$$R_{3r} = \frac{-1}{8 \cdot \pi \cdot (r_a^2 - r_i^2) \cdot \lambda_r \cdot l_z}\left[(r_a^2 + r_i^2) - \frac{4 \cdot r_a^2 \cdot r_i^2 \cdot ln(\frac{r_a}{r_i})}{r_a^2 - r_i^2}\right] \tag{8.11}$$

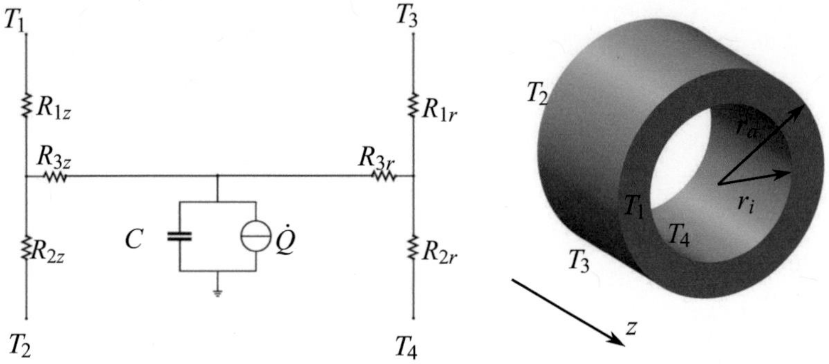

Figure 8.22. Equivalent thermal network of a hollow cylinder

Hollow cylinder segment

The equivalent thermal network of a hollow cylinder segment consists of three T-shape equivalent circuits, as shown in Fig. 8.23.

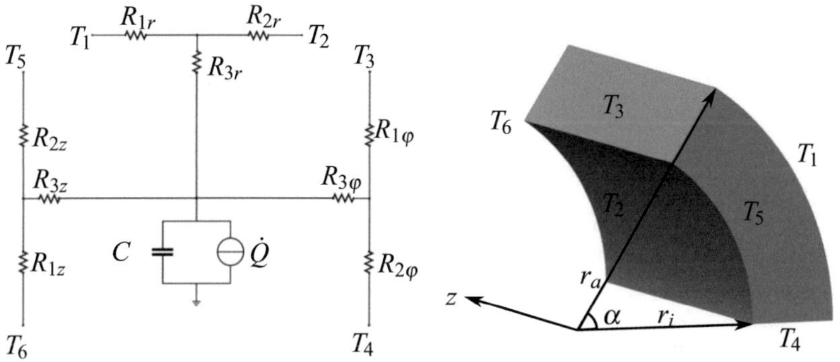

Figure 8.23. Equivalent thermal network of a hollow cylinder segment

The resistances corresponding to the axial direction can be calculated with the same method as for the whole hollow cylinder, while the resistances relevant to the radial direction must be modified considering the reduced area.

$$R_{1r} = \frac{90^\circ}{\alpha \cdot \pi \cdot \lambda_r \cdot l}\left[1 - \frac{2 \cdot r_i^2 \cdot ln(\frac{r_a}{r_i})}{r_a^2 - r_i^2}\right] \tag{8.12}$$

$$R_{2r} = \frac{90^\circ}{\alpha \cdot \pi \cdot \lambda_r \cdot l}\left[\frac{2 \cdot r_a^2 \cdot ln(\frac{r_a}{r_i})}{r_a^2 - r_i^2} - 1\right] \tag{8.13}$$

$$R_{3r} = \frac{-45^\circ}{\alpha \cdot \pi \cdot (r_a^2 - r_i^2) \cdot \lambda_r \cdot l}\left[(r_a^2 + r_i^2) - \frac{4 \cdot r_a^2 \cdot r_i^2 \cdot ln(\frac{r_a}{r_i})}{r_a^2 - r_i^2}\right] \tag{8.14}$$

In order to calculate the resistances related to the tangential direction, the equivalent length should be firstly defined.

$$l_\varphi = \frac{\alpha}{360^\circ} \cdot \pi \cdot (r_i + r_a) \tag{8.15}$$

$$R_{1\varphi} = \frac{l_\varphi}{2 \cdot \lambda_\phi \cdot (r_2 - r_1) \cdot z} \quad (8.16)$$

$$R_{2\varphi} = \frac{l_\varphi}{2 \cdot \lambda_\phi \cdot (r_2 - r_1) \cdot z} \quad (8.17)$$

$$R_{3\varphi} = -\frac{l_\varphi}{6 \cdot \lambda_\phi \cdot (r_2 - r_1) \cdot z} \quad (8.18)$$

The other mechanism considered in the modelling is the convection, which describes the thermal exchange between the solid and the surrounding fluid through advection or diffusion or a combination of both. In general, the exchanged heat flux can be determined using (8.19), where A and α are the contact area and the convective heat transfer coefficient.

$$\dot{Q} = A \cdot \alpha \cdot \Delta T \quad (8.19)$$

Furthermore, the convective heat transfer coefficient α can be determined using (8.20) with the dimensionless Nusselt number N_u, the thermal conductivity λ and the characteristic length L.

$$\alpha = \frac{\lambda}{L} \cdot N_u \quad (8.20)$$

The calculation of Nusselt number is very complicated, which depends on the type of the flow and the specific geometry.

Natural convection

For the natural convection, the fluid movements due to the different density within the fluid resulted from the inhomogeneous temperature. In this case, the Nusselt number is a function of the Grashof number G_r, the Prandtl number P_r and the geometry.

$$N_u = f(G_r, P_r, geometry)$$

The Grashof number is defined as the ratio of the buoyancy to viscous force acting on the fluid and can be calculated using (8.4.2), where β and ν are the thermal expansion coefficient and the kinematic viscosity of fluid.

$$G_r = \frac{\beta \cdot g \cdot L^3 \cdot \Delta T}{\nu^2}$$

The Prandtl number defined as the ratio of momentum diffusivity to thermal diffusivity can be calculated with (8.4.2), where a and c_p are the thermal diffusivity and the specific heat capacity.

$$P_r = \frac{\nu}{a} = \frac{\nu \cdot \rho \cdot c_p}{\lambda}$$

Forced convection

In contrast to the natural convection, the movement of the fluid for forced convection is caused by the pressure difference resulted by the external force. In this case, the Nusselt number is accordingly depending on the Reynolds number R_e, the Prandtl number P_r and the specific geometry.

$$N_u = f(R_e, P_r, geometry)$$

The Reynolds number R_e is often used to characterize the flow regimes with in a fluid, such as laminar or turbulent flow. It is defined as the ratio of momentum forces to viscous forces using (8.4.2), where the v is the relative velocity between the object and fluid and μ is the dynamic viscosity of the fluid.

$$R_e = \frac{\rho \cdot v \cdot L}{\mu} = v \cdot L\nu$$

Combined natural and forced convection

In the praxis, the natural convection is often superimposed by forced convection. In this circumstance, the effective Nusselt number can be calculated depending on the direction of the fluid movement resulted by the natural and forced convection respectively.

$$N_u = \begin{cases} \sqrt[3]{N_{u,force}^3 + N_{u,natural}^3} & \text{opposite} \\ \sqrt[3]{N_{u,force}^3 + N_{u,natural}^3} & \text{parallel} \end{cases}$$

Based on above analysis, the Nusselt number N_u or rather convective heat transfer coefficient α of the surfaces of solid components can be determined according to the air flow velocity and position. For instance, some typical surfaces in Fig. 8.24 are marked with different colours according to the type of convection.

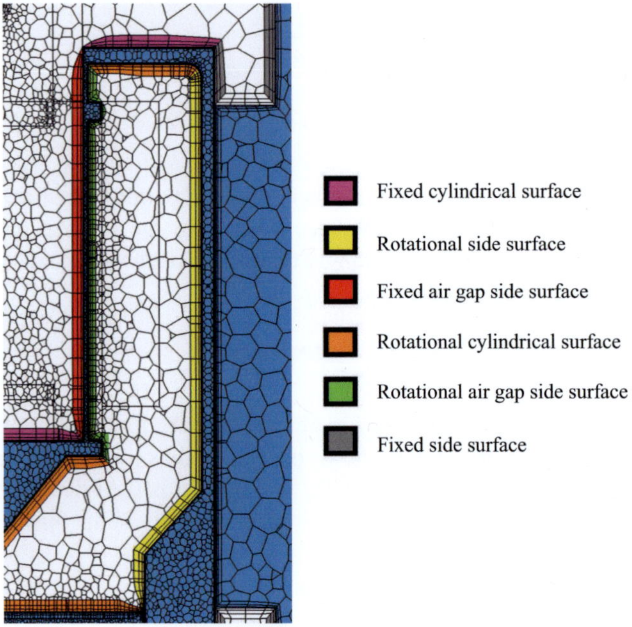

Figure 8.24. Different types of convection surfaces

The thermal network of the complete PMAFM-ISA has been developed, which is illustrated in Fig. 8.25. Each of the coloured block corresponds to a component of the prototype. For instance, the equivalent thermal network of the PM mounted on the rotor can be clearly seen in Fig. 8.25, which consists of one thermal source corresponding to the eddy current losses in PM, one block considering the convective heat exchange and thermal network of hollow cylinder segments consisting of three T-shape equivalent circuits.

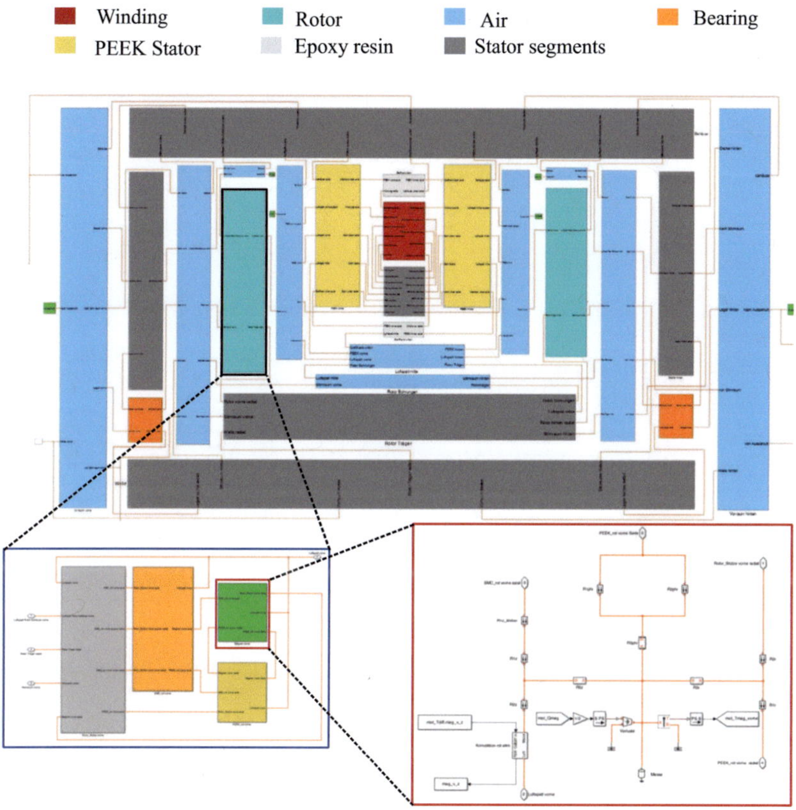

Figure 8.25. Equivalent thermal network of the PMAFM-ISA

The analysis process is completely controlled with scripts. After all geometric parameters and losses are imported, the simulation is performed. At last, the temperature of all components are saved and the same process is repeated until the results for all operating points are obtained.

8.4.3 Results of the analytical thermal analysis

In order to validate the analytical model, the determined temperature of some critical components are compared with that achieved from the coupled thermal and CFD analysis for 24 representative operating points covering almost the complete efficiency map of PMAFM-ISA.

The operating points correspond to the rotational speed $200\,\text{min}^{-1}$, $500\,\text{min}^{-1}$, $1400\,\text{min}^{-1}$, $1900\,\text{min}^{-1}$, $2600\,\text{min}^{-1}$, $4500\,\text{min}^{-1}$ and 0%, 10%, 40%, 70% and 100% of the maximum torque for each speed. For instance, the results the operating points corresponding to $1400\,\text{min}^{-1}$ are listed in Table 8.10, where the position 1 is defined as the position under the middle of the stator bar.

Torque ratio	Magnet		Pos. 1	
	Ana.	Num.	Ana.	Num.
0%	67.89 °C	68.15 °C	55.71 °C	56.58 °C
10%	56.04 °C	60.14 °C	61.38 °C	63.17 °C
40%	71.52 °C	76.96 °C	91.6 °C	93.79 °C
70%	105.85 °C	114.23 °C	157.58 °C	160.72 °C
100%	168.44 °C	177.7 °C	261.83 °C	265.44 °C

Table 8.10. Temperature of prototype at $1400\,\text{min}^{-1}$ based on analytical and numerical calculation

It can be noted that the analytical model can achieve relatively high accuracy, which can therefore be utilized to calculate the temperature field based on the analysis of all operating points within the complete efficiency map.

Based on the electromagnetic design PMAFM-ISA1 and the mechanical construction, the average temperatures of the stator segment and PM for the complete efficiency map can be determined, as illustrated in Fig. 8.26.

It can be noted that a very high stationary temperature is achieved in both stator segment and PM, especially in the high speed area. Considering that the maximum allowable temperature of stator segment and PM are equal to 180 °C and 120 °C respectively, it is clear that the continuous power of PMAFM-SAT1 is very limited.

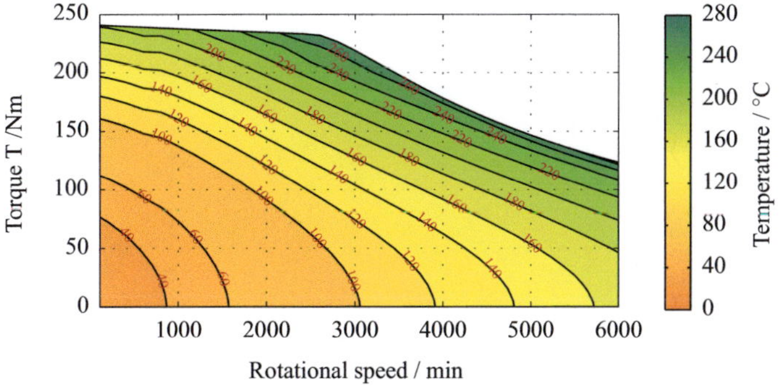

(a) Stator segment

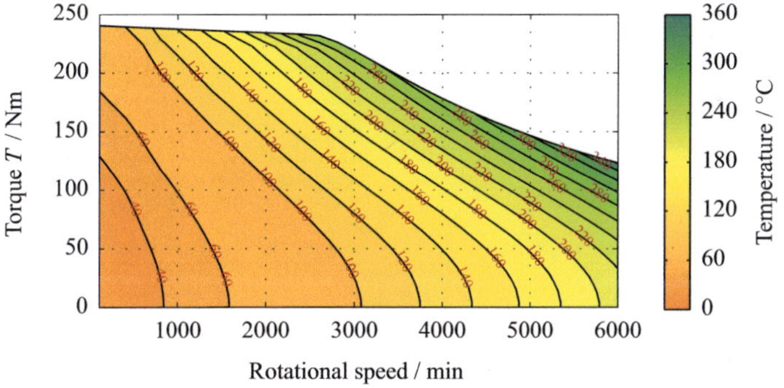

(b) PM

Figure 8.26. Temperature of PMAFM-ISA1 based on analytical model

In comparison, the stationary temperature field of the stator segment and PM of the PMAFM-ISA2 with PM segmentation are illustrated in Fig. 8.27. It can be noted that the temperature of both stator segment and the PM reduces significantly. In other words, the continuous power of PMAFM-SAT2 can be obviously improved.

In this chapter, a possible mechanical construction for PMAFM-ISA is developed based on the results of electromagnetic analysis, which is then validated with FEM in terms of the stress and deformation of all critical components. In addition, a test to investigate the performance of adhesive at different temperature has also been performed.

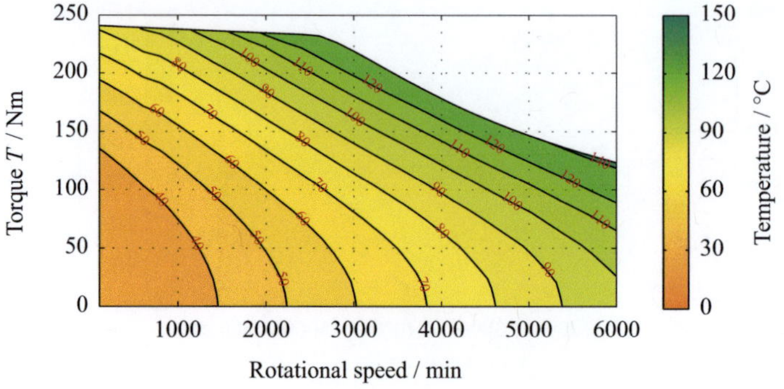

(a) Stator segment

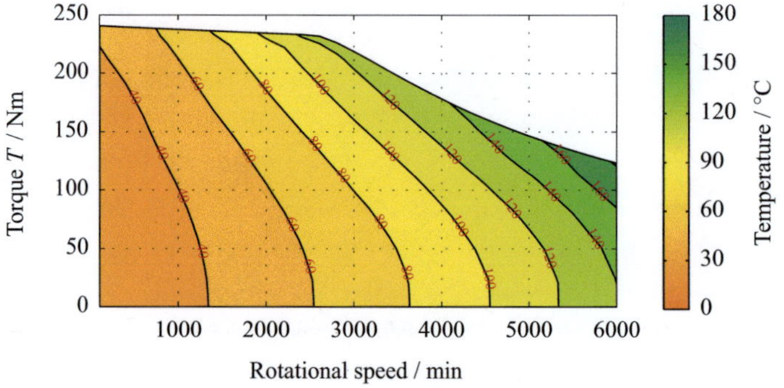

(b) PM

Figure 8.27. Temperature of PMAFM-ISA2 based on analytical model

In order to determine air flow and temperature of PMAFM-ISA, coupled thermal and fluid dynamic analysis has been carried out. In order to reduce computational time, an analytical model has been developed as well. From the calculated temperature field, it is clear that for PMAFM-ISA the stationary temperature of both stator segment and PM reduces significantly through the segmentation of PM, which leads to an obvious increase of the continuous power.

9

Prototyping and Measurement

In addition to the adhesive test described in the chapter 8, the following four preliminary tests have been performed before the manufacturing of the components as well, as illustrated in Fig. 9.1.

- Winding test

  With the help of the winding test, the round and flat wires with different sizes are tested. Finally, the flat wire 3x1mm is selected for the prototyping.

- PEEK machining test

  Considering the few experience with machining of PEEK material, the test is necessary to determine suitable machining method and machining accuracy.

- Direct current test

  Purpose of the test is to validate the stability of the connection between the copper ring and the bar connector illustrated in Fig. 8.5 at high temperature, which is of great importance for the stability of the electrical system at high electrical loading.

- Casting test

  All components of stator such as SMC segments with concentrated winding, the electrical system, the PEEK components, the water cooling system and so on are cast together with the epoxy resin to improve the mechanical stability and to reduce the thermal resistance. With this test, the performance of the used epoxy resin is tested, especially the viscosity and the curing temperature.

(a) Winding test

(b) PEEK machining test

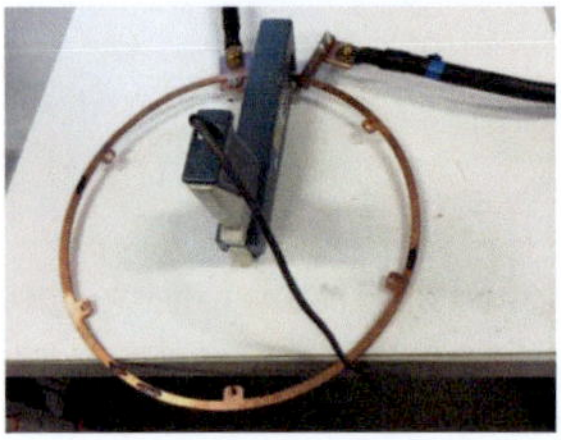

(c) Direct current test

(d) Casting test

Figure 9.1. Preliminary tests as preparation of prototyping

Based on the previously developed mechanical construction, the detailed electromagnetic and mechanical analyses as well as the performed preliminary tests, the prototyping and the measurement of the prototype have been carried out and described in this chapter. At first, the technical drawings of each customized components are created. Together with the standard components such as resolver, the stator and the rotor are assembled and tested respectively. On this basis, the prototype of PMAFM-ISA1 has been built up and measured on the test bench.

9.1 Prototyping

In this section, the manufacturing of the most crucial components are firstly presented. Subsequently, the assembly and tests of the stator and the rotor as well as the assembly of the complete prototype are described.

9.1.1 Manufacturing of customized components

Compared with the standard components such as resolver, the customized components should be prepared with more attention. Hereby, the pretreatment of SMC segments, the manufacturing of PEEK plates, the non-standard steel and aluminium components are presented.

Stator segments

As mentioned above, all components of stator are cast together to improve the mechanical firmness and to reduce the thermal resistance. Considering the high potential risk, the casting process is divided into two steps instead of casting all components once.

Firstly, all stator segments with the concentrated winding are cast with a casting mould. Subsequently, the examined stator segments are cast together with other components of stator. The casting mould illustrated in Fig. 9.2 comprises 12 steel plates with different thickness.

Figure 9.2. Assembled casting form and steel plates

The pretreatment of the stator segment consists mainly of six steps and some of them are illustrated in Fig. 9.3. Considering the time-consuming curing process of epoxy resin, there are totally four casting moulds manufactured in order to cast the 36 stator segments within one month.

- SMC segment is wound with flat wire to achieve required number of turns.
- Wound SMC tooth is mounted and fixed in the casting mould.

- Epoxy resin should be prepared by mixing two components evenly according to the specified ratio.
- Mould with embedded wound segment is cast with prepared epoxy resin.
- Mould is placed into the oven and heated until the epoxy resin is cured.
- Mould is disassembled and cleaned for the next casting.
- Cast segment is cleaned and examined by measuring the electrical resistance.

(a) Winding of segment (b) Casting mould with inserted segment (c) Cured resin (d) Cleaned tooth

Figure 9.3. Manufacturing process of stator segment

PEEK plates

Several important components in the prototype are made from PEEK, which have annular form and aim to fix the stator segments and PM accurately. Initially, the PEEK plates are manufactured with water jetting. However, the required manufacturing precision cannot be achieved and the cutting edge isn't perpendicular to the surface of the plates. Besides, unwanted cracks and chipping have been found during the machining process.

Therefore, the PEEK plates are manufactured with the CNC-milling machine. The raw material of PEEK plate is clamped with a vacuum table and milled to the required thickness. In order to limit the deformation during the machining, the PEEK plate is milled only for maximum 0.3 mm at each time and the machining sides are alternately changed. The machining process and the manufactured PEEK plates are illustrated in Fig. 9.4.

(a) Fixation of PEEK with vacuum table

(b) Milled PEEK plate with required thickness

(c) PEEK plate of stator

(d) PEEK plate of rotor

Figure 9.4. Machining process of PEEK plates

Steel and aluminium parts

The technical drawings are the basis of the manufacturing of the customized steel and aluminium parts, where the dimensional, positional and geometrical tolerances as well as the fitting are specified. Therefore, the technical drawings are of great importance for the functionality and stability of the prototype.

For instance, the main shaft made from 16MnCr5 and its manufacturing drawings are illustrated in Fig. 9.5. It can be noted that a large number of tolerances are specified, especially for the surfaces used to connect the bearings and the shaft sleeve.

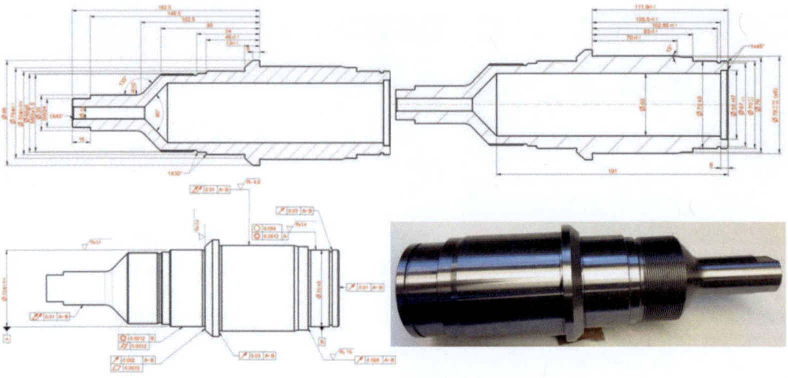

Figure 9.5. Technical drawing of the main shaft

9.1.2 Assembly

Besides the above manufactured customized components, the other standard components are ordered as well. In this subsection, the comprehensive assemblies of the stator, the rotor and the complete prototype are described.

Stator

Each phase of PMAFM-ISA consists of 6 parallel paths and in each path there are two serially connected segments. The sorted segments belonging to a path are crimped together, as illustrated in Fig. 9.6. Therefore, the previously cast and examined stator segments should be connected so that the resistance of each phase should be as equal as possible.

Subsequently, all the stator segments are connected with the electrical system, which are firmly fixed by the PEEK plates and cast together with other components of stator. The assembled components of the stator and the cast stator are illustrated in Fig. 9.7.

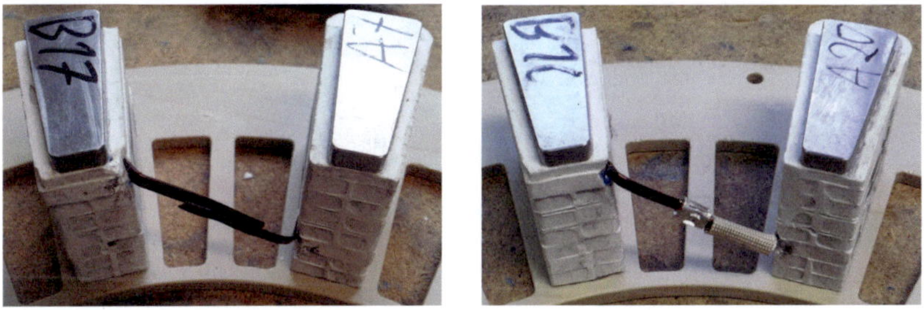

(a) Two isolate segments (b) Connected segments

Figure 9.6. Isolate and serially connected segments

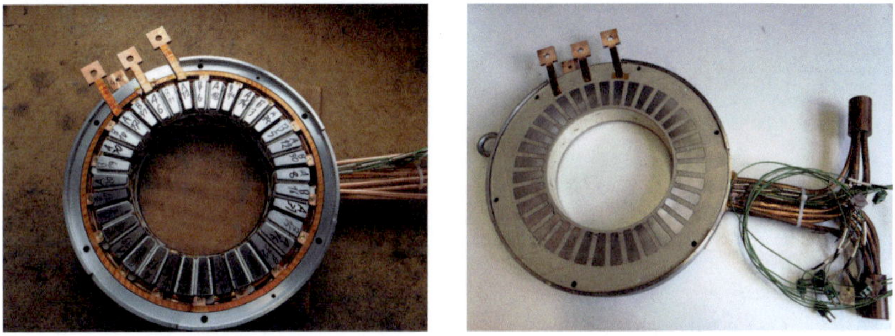

(a) Before casting (b) After casting

Figure 9.7. Stator before and after casting

At last, the cast stator should be examined before the assembly of complete prototype. Firstly, the high voltage isolation test and the water seal test are carried out. Subsequently, the resistances of three phases are measured and compared, as listed in Table 9.1. It can be noted that the differences among the phases are relatively small, which leads to a homogeneous distribution of the current from inverter.

	Phase U	Phase V	Phase W
Resistance [mΩ]	6.238	6.233	6.235

Table 9.1. Phase resistances of the PMAFM-ISA

Rotor

In comparison, the assembly of the rotor is much easier. First of all, the PMs and the rotor yoke as well as the rotor carrier are glued together with the industrial adhesive, which is then placed at at oven and heated to cure the adhesive. On the other side, the shaft sleeve is heated to the temperature listed in Table 8.5 and connected with the shaft through shrinking, as shown in Fig. 9.8.

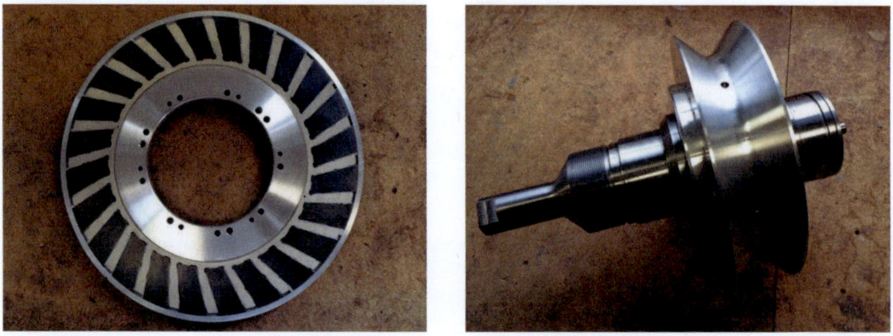

(a) Glued components of rotor
(b) Shrunk shaft and shaft sleeve

Figure 9.8. Partly assembled Rotor

Finally, the two separate parts of rotor are shrunk and additionally screwed together for a better stability during the operation. Subsequently, the manufactured rotor has been balanced to distribute the mass evenly around the rotational axis, which leads to reduction of the noise and vibration as well as improvement of bearing life, as illustrated in Fig. 9.9.

(a) Assembled rotor

(b) Balanced rotor

Figure 9.9. Assembled and balanced rotor

Complete prototype

Based on the above work, the complete prototype is assembled. At first, the bearings are mounted on the main shaft through shrinking. Subsequently, the housing and the stator are screwed together. Finally, the rest components such as the resolver and the cover are mounted on the prototype, which is illustrated in Fig. 9.10.

(a) Real side

(b) Front side

Figure 9.10. Completely assembled prototype

9.2 Measurement of the prototype

The assembled prototype should be measured to investigate the performance of PMAFM-ISA and to validate the FEM results. In this section, the test stand used to measure the prototype is firstly introduced. Subsequently, the measuring results and the comparison with the FEM results are presented.

9.2.1 Test stand

The measurement of the prototype has been carried out in a test stand developed by ETI, which aims to fully characterize different kinds of electrical machines in a large power spectrum. With its help, the performed simulation including electromagnetic and thermal analyses can be validated. The test stand and the main components as well as the mounted prototype are illustrated in Fig. 9.11.

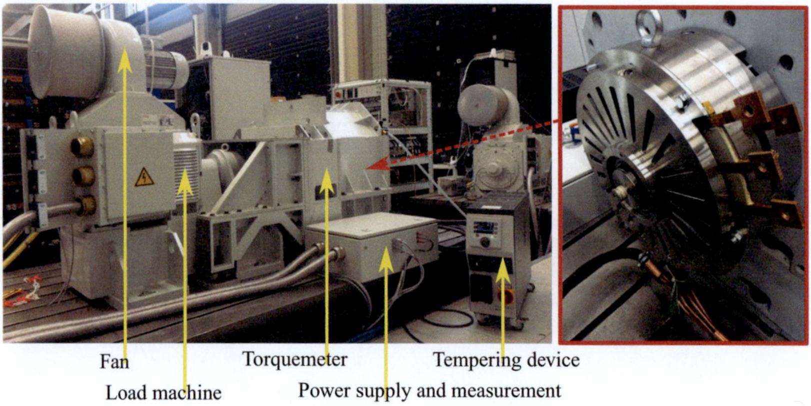

Figure 9.11. Test stand with mounted prototype

An asynchronous machine (ASM) from the company OSWALD is utilized as load machine, whose rotational speed is controlled by a digital signal processor (DSP). The losses and efficiency of the prototype are measured with an accurate power analyser WT3000 of the company YOKOGAWA based on measured data of the rotational speed, torque, current and voltage of the three phases.

The phase current is measured with the current transducers of the company LEM and evaluated with the electronics developed by the company SIGNALTEC. In addition, a temperature control device is connected with the cooling system of the prototype to adjust the temperature and to dissipate the generated heat.

The torque is measured with the non-contact measuring system from the company MANNER, which consists of a measuring flanges, a telemetry and a evaluating unit. The rotational speed is measured with an optical incremental encoders, while the measurement of torque is performed with a strain gage.

In addition, some thermometers are cast in the stator to measure and monitor the temperature of the stator. The signals of the thermometers are evaluated with the system of the company HBM and further transmitted to the central control unit.

The schematic diagram of the test stand is shown in Fig. 9.12, which comprises two AC-DC converters and two DC-AC converters. The intermediate circuit

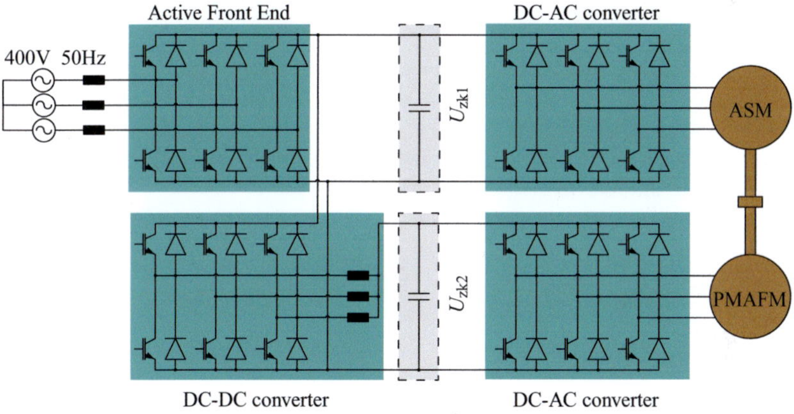

Figure 9.12. Schematic diagram of the test stand

voltage U_{zk1} is provided with an AC-DC converter (Active Front End-AFE), which enables power supply with few harmonics, free of reactive power and the ability of returning energy to the grid. With a DC-AC converter, the three-phase AC-voltage can be generated and supplied to the load machine. Besides, the required intermediate circuit voltage U_{zk2} can be controlled with a three phase DC-DC converter, which is transformed into three-phase AC-voltage with the following DC-AC converter to feed the prototype.

A main advantage of this structure is that the power delivered by an electrical machine is absorbed by the other and flows in the circuit, so that only the power losses in the components of test stand or the dynamical power requirement should be covered by the grid. Another merit of the structure with two independent intermediate circuit voltages U_{zk1} and U_{zk2} is that a complete system in an electrical vehicle including battery, converter and driving machine can be simulated, which can offer more convincing and practical results.

9.2.2 Measurement of prototype with no-load

First of all, the performance of the prototype without electrical loading is measured. Subsequently, the results are compared with the 3D FEM results from the electromagnetic and thermal analyses.

No-load losses

A comparison of the no-load losses is illustrated in Fig. 9.13. It should be noted that the losses of simulation is made up of the eddy current losses of PM and the iron losses of SMC, while the measured results consist of the friction losses resulted from air and bearings additionally.

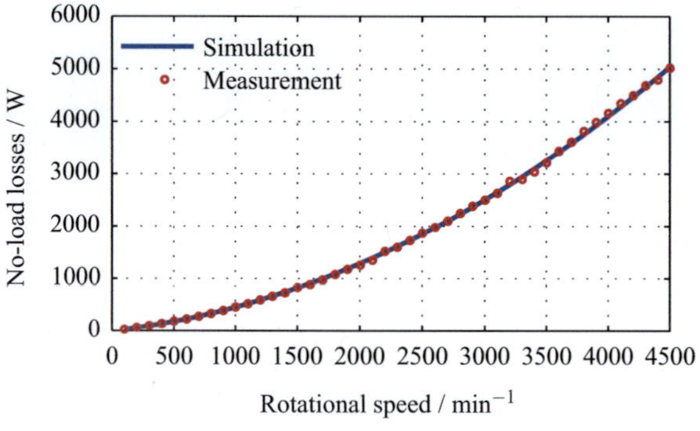

Figure 9.13. Comparison of the no-load losses

Apparently, the deviations between the simulation and measurement are extraordinarily small, which implies the iron losses are accurately calculated with the developed model. Furthermore, the losses resulted by friction is relatively small and even negligible.

No-load induced voltage

Subsequently, the induced voltage of the prototype at no-load condition is measured at different rotational speed, which is of great importance for the performance of an electrical machine.

For instance, the induced voltage at 200 min^{-1} is illustrated in Fig. 9.14. It is clear that the induced voltage of the three phases are almost the same, except for the shift corresponding to the electrical angle of 120°.

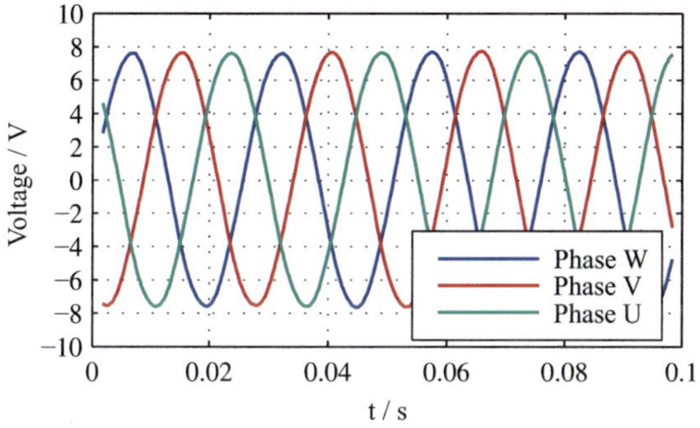

Figure 9.14. Measured induced voltage of the prototype at 200 min^{-1}

Furthermore, a comparison of the induced voltage based on the simulation and measurements is illustrated in Fig. 9.15. It can be noted the simulation agrees well with the measurement in the low speed area and the deviation becomes larger with increasing speed. A reason is that the eddy currents induced in the PMs at high rotational speed are underestimated, which can reduce the PM flux flowing into the stator segments.

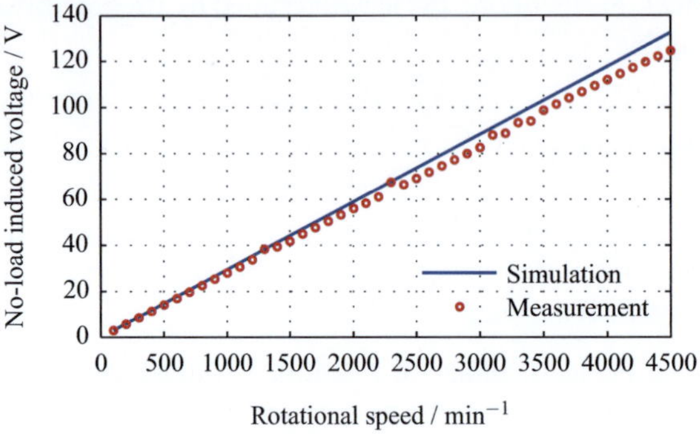

Figure 9.15. Comparison of the simulated and measured no-load induced voltage

Cogging torque

The cogging torque is an important part of the torque ripple of an electrical machine during the operation, which can lead to serious vibration and noise. For this reason, great attentions have been paid to the skewing of the PM during the previous electromagnetic design. A comparison of the simulated and measured cogging torque has been carried out, as illustrated in Fig. 9.16.

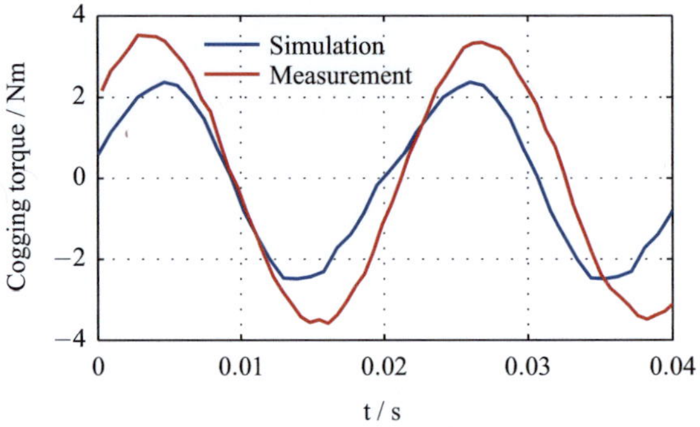

Figure 9.16. Comparison of the simulated and measured cogging torque

As shown in Fig. 9.16, the maximum deviation is about 1.15 Nm. However, it should be noted that the measured cogging torque consists not only the torque generated through the interaction between the rotor and the stator, but also the torque resulted from the friction of the bearings in the load machine and the prototype. For this reason, it can be concluded that the cogging torque is relatively accurately calculated.

Temperature of prototype at no-load

The calculation of the temperature based on the coupled thermal and fluid dynamical analysis is presented in the last chapter. A comparison of the temperature of the prototype without electrical loading is more significant as the thermal sources, namely the no-load loss, is more accurately calculated than the circumstance that the machine is fed with an inverter.

The results for PMAFM-ISA at different operating points are compared and listed in Table 9.2, where the position 1 corresponds to the position in the inner radius of the stator, while position 2 is defined as the position between the stator segments at the average radius of the stator.

Speed	Numerical		Measurement	
	Position1	Position2	Position1	Position2
1000 min^{-1}	45.13 °C	43.12 °C	43.92 °C	42.96 °C
1400 min^{-1}	56.58 °C	53.78 °C	56.08 °C	54.81 °C
1900 min^{-1}	71.34 °C	67.61 °C	68.70 °C	67.42 °C

Table 9.2. Comparison of the simulated and measured temperature

It can be noted that the results based on the coupled thermal and fluid dynamic analysis match well with the measurement. The maximum error occurs at rotational speed 1900 min^{-1}, which is equal to 1.64 °C for position 1 and 0.19 °C for position 2. The deviation is resulted from many reasons and some of them are:

- The free space within the stator isn't completely filled with the epoxy resin. There might be small air pockets existing in the stator, which reduces the thermal conductivity and results in a higher temperature.
- There might be small gaps between the concentrated winding and the stator segments, which cannot be measured and taken into account in FEM.
- The temperature is measured with the thermometers, whose positions might

be slightly changed during the casting. Considering the high temperature gradient in the stator, it might have great influence on the measured results.

9.2.3 Measurement of prototype with electrical loading

The measurement of the prototype is much complicated than that at no-load and the results are strongly depending on the current controlling quality. Afterwards, the modulation method of the DC-AC inverter and the current controller of the prototype are described. The control of AFE, the DC-DC converter and the DC-AC converter feeding the ASM aren't the focuses of this work and will not be explained.

Modulation of the DC-AC inverter

As illustrated in Fig. 9.17, the DC-AC inverter consists of six diodes and six insulated gate bipolar transistors (IGBTs), which can be switched on and off by gate signals. On the other hand, each phase of an electrical machine is modelled as a series connection of RL-load and countervoltage. The IGBTs in a half bridge (HB) should be inversely switched to avoid the short circuit.

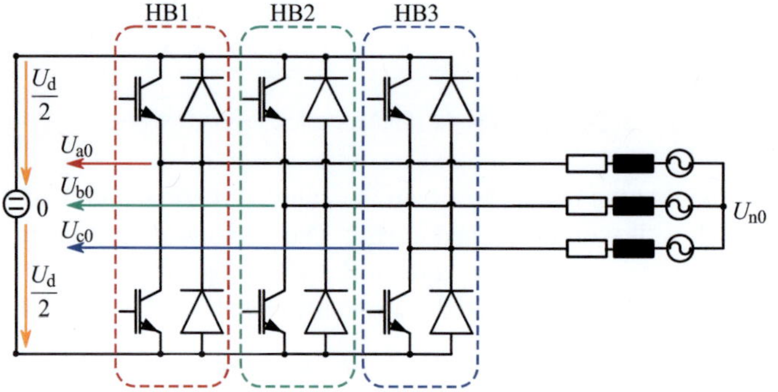

Figure 9.17. Structure of a DC-AC inverter

There are many different methods to control the switching of IGBTs and to generate the wanted phase voltage from DC voltage U_d such as sinusoidal pulse width modulation (SPWM), the super SPWM, the flat-top-modulation and the space vector modulation.

The method utilized in this work is the super SPWM, where the controlling signals of IGBTs are generated by comparing the wanted sinusoidal signals with a triangular signal, which has normally obviously higher frequency than the wanted sinusoidal signal.

Compared with the traditional SPWM method, an increase by nearly 15 percent of the phase-to-neutral voltage U_{an}, U_{bn} and U_{cn} can be achieved with the super SPWM by adding a voltage component U^{*}_{n0} in the desired phase voltage U^{*}_{a0}, U^{*}_{b0} and U^{*}_{c0}, as illustrated in Fig. 9.18.

$$U^{*}_{\mathrm{n0}} = -\frac{\max(U^{*}_{\mathrm{a0}},U^{*}_{\mathrm{b0}},U^{*}_{\mathrm{c0}}) + \min(U^{*}_{\mathrm{a0}},U^{*}_{\mathrm{b0}},U^{*}_{\mathrm{c0}})}{2}$$

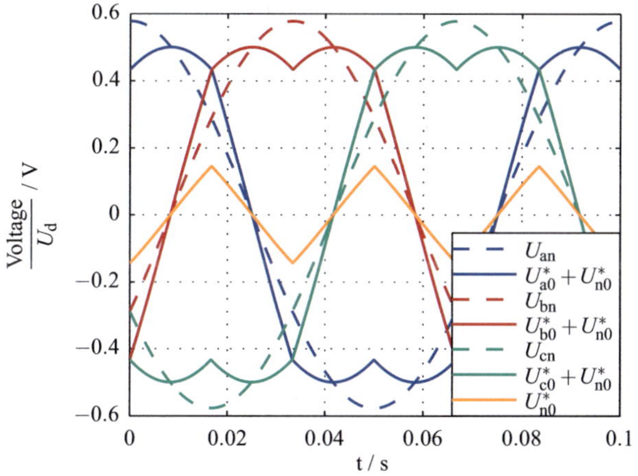

Figure 9.18. Phase voltage and Phase-to-neutral voltage of SPWM

Current controlling

During the measurement, the current i_{d} and i_{q} should be controlled to the desired values, which can be determined through the Clarke-transformation using (9.2.3), where $\underline{a}$ is equal to $e^{j2\pi/3}$.

$$i_{\mathrm{d,q}} = \frac{2}{3} \cdot (i_{\mathrm{a}} + \underline{a} \cdot i_{\mathrm{b}} + \underline{a}^{2} \cdot i_{\mathrm{c}}) \cdot e^{-j\gamma_{el}}$$

The i_a and i_b and i_c are the phase currents, while γ_{el} is the the electrical angle of the rotor measured by the resolver.

As illustrated in Fig. 9.19, the controlling of i_d and i_q is achieved with two PI-controllers. Besides, the decoupling is integrated to realize a separate controlling of the two current components.

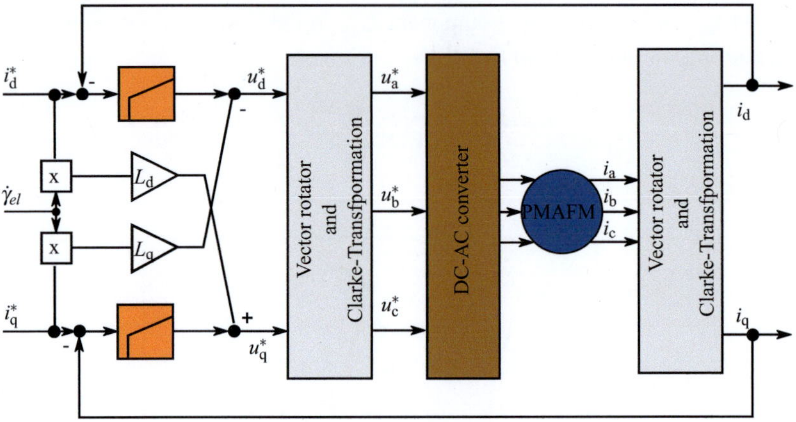

Figure 9.19. Current control of the prototype

Measuring routine

The prototype should be measured for a large number of operating points of the prototype at different speed, which correspond to many combinations of current components. For this reason, a measuring routine becomes necessary to perform a specified measuring process and to save the results during the measurement automatically.

The measuring routine is an interface between the PC and the central control unit, namely the digital signal processor(DSP). It is developed with the software LabVIEW of the company NATIONAL INSTRUMENTS, whose flow chart is illustrated in Fig. 9.20.

At first, the parameters including the rotational speed, the current components i_d and i_q are specified. Subsequently, the temperature of the stator is examined. To eliminate the influence of the temperature, the measurement is carried out only if the temperature is within the tolerance range. Besides, the thermal conditioning is realized by controlling the current i_d and i_q. If all previously specified parameters are achieved or within the tolerance range, the values of required parameters are then saved. The process goes further until all required operating points are measured.

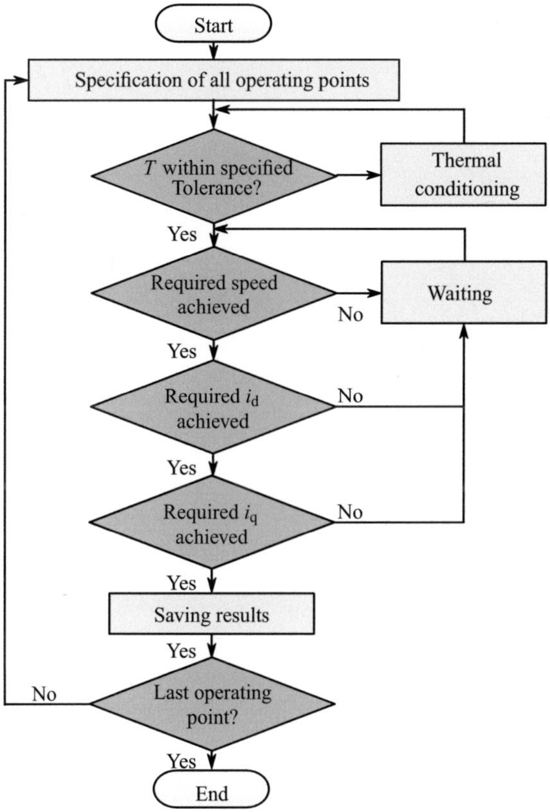

Figure 9.20. Flow chart of the measuring routine

Measurement of torque

The torque is one of the most important properties for electrical machine. For instance, the torque of the PMAFM-ISA at rotational speed equal to $200\,\mathrm{min}^{-1}$ and $2000\,\mathrm{min}^{-1}$ with $i_\mathrm{d} = 0A$ and different i_q is illustrated in Fig. 9.21.

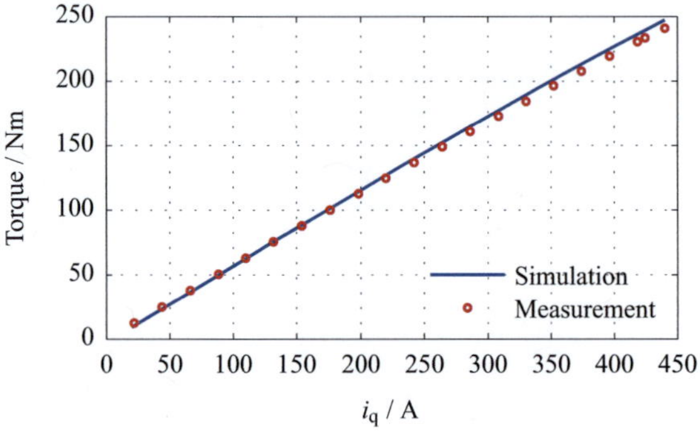

(a) Speed $200\,\mathrm{min}^{-1}$

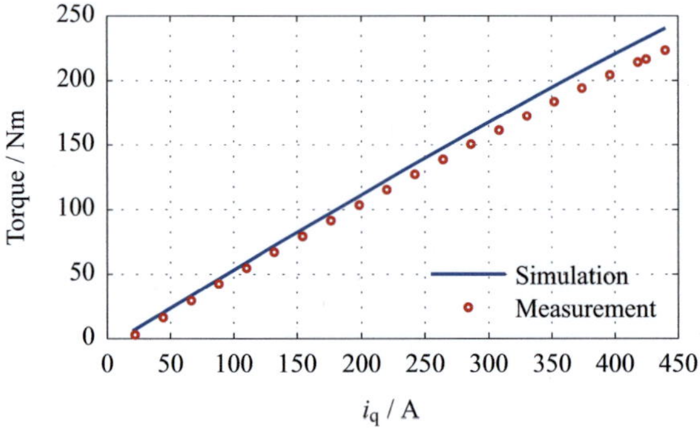

(b) Speed $2000\,\mathrm{min}^{-1}$

Figure 9.21. Comparison of the torque for $i_\mathrm{d} = 0\,\mathrm{A}$

For the maximum current $i_\mathrm{q} = 440\,\mathrm{A}$, there is a torque deviation equal to 6.39 Nm for rotational speed $200\,\mathrm{min}^{-1}$, while the deviation becomes larger, which equals 17.1 Nm for rotational speed $2000\,\mathrm{min}^{-1}$.

It should be noted that the torque measurement should be carried out for all possible current component combinations and different rotational speed. For example, the measured output torque of the prototype for all the operating points at 500 min^{-1} is illustrated in Fig. 9.22.

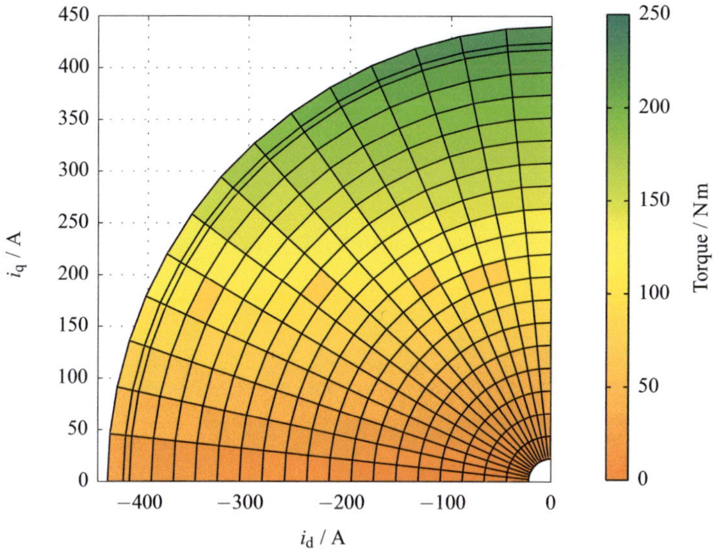

Figure 9.22. Torque measured at 500 min^{-1}

The controlling quality of current is the most important factor, which causes the deviation illustrated in Fig. 9.21. As illustrated in Fig. 9.23, the actual current value varies continuously around the required i_d and i_q, which should be constant 0 A and 440 A respectively.

Because of the high pole pair number and the skewing of PM, the controlling quality of current becomes obviously poorer when the prototype rotates at high speed. In the measurement, many harmonics exist in the current, while the i_d and i_q are ideally constant in FEM.

As a result, more eddy current losses in PMs and more iron losses in SMC are generated, which result in reduction of the available output torque. Furthermore, the additionally generated losses lead to fast heating of the adhesive and reduction of the continuous output power.

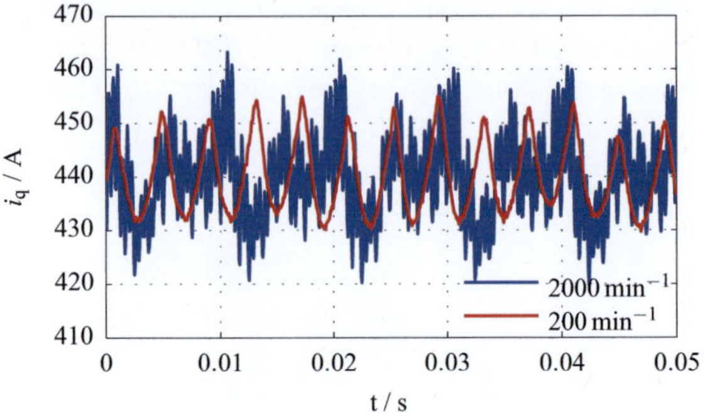

(a) Measured current i_q

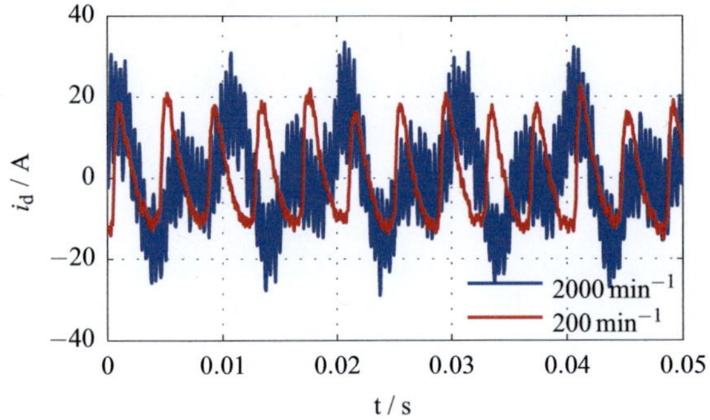

(b) Measured current i_d

Figure 9.23. Measured current i_d and i_q at 200 min^{-1} and 2000 min^{-1}

Measurement of efficiency

Besides the torque, the efficiency is another important property of the prototype, which is defined as the ratio between the mechanical output power and the input electrical power. For instance, the efficiency and losses of the prototype at 2000 min^{-1} with $i_d = 0$ A and different i_q are illustrated in Fig. 9.24.

It can be noted that for the rotational speed 2000 min^{-1}, the efficiency deviation between the simulation and measurement increases with higher current i_q. The maximum deviation equals nearly 2 %. The most important reason is that the total losses deviation increases dramatically due to the poor current controlling quality as illustrated in Fig. 9.23.

With increasing i_q, more harmonics are contained in current, which lead to higher copper losses and lower efficiency. On the other hand, the deviation of efficiency raises slightly due to increased temperature during the measurement.

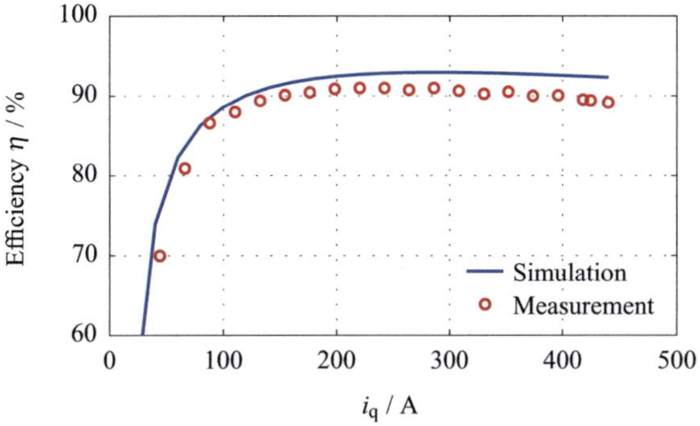

(a) Efficiency

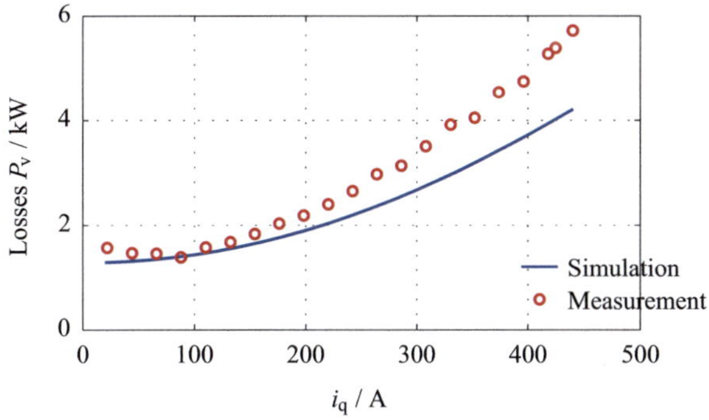

(b) Total losses

Figure 9.24. Simulated and measured efficiency and losses for prototype with $i_d = 0\,\mathrm{A}$ at $2000\,\mathrm{min}^{-1}$

Measurement of flux linkage

The flux linkage Ψ_d and Ψ_q can be determined from the measured stationary results of the prototype. For this case, the flux linkage can can be calculated as follows.

$$\Psi_\mathrm{q} = \frac{R_\mathrm{S} \cdot i_\mathrm{d} - u_\mathrm{d}}{\omega_\mathrm{el}} \tag{9.1}$$

$$\Psi_\mathrm{d} = \frac{u_\mathrm{q} - R_\mathrm{S} \cdot i_\mathrm{q}}{\omega_\mathrm{el}} \tag{9.2}$$

The calculated flux linkage Ψ_d and Ψ_q based on the measurement at 500 min^{-1} are illustrated in Fig. 9.25 and Fig. 9.26 respectively.

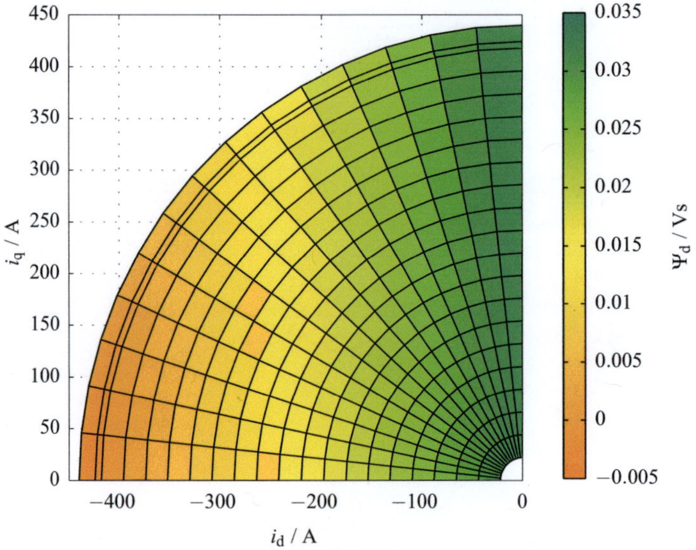

Figure 9.25. Measured flux linkage Ψ_d at 500 min^{-1}

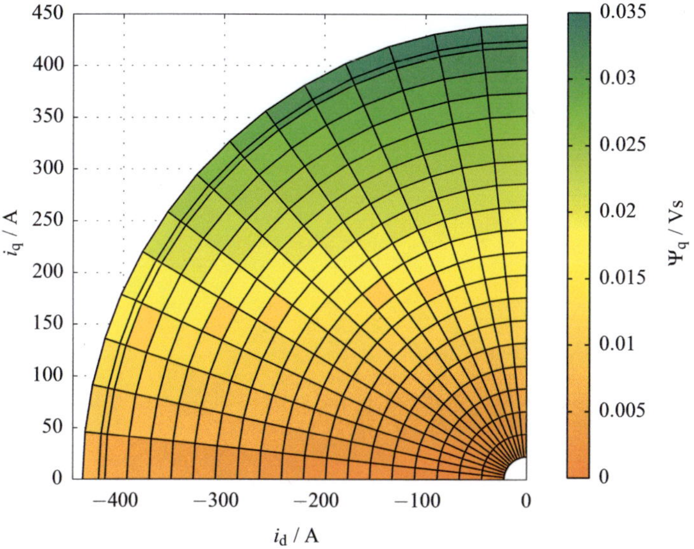

Figure 9.26. Measured flux linkage Ψ_q at $500\,\mathrm{min}^{-1}$

In comparison, the flux linkages Ψ_d and Ψ_q determined with the 3D FEM are illustrated in Fig. 9.27. It is clear that the FEM results match very well with the measured flux linkage.

Measurement of the limiting characteristic curve

The limiting characteristic curve presents the maximum output torque and power of an electrical machine at different rotational speed, which determines the boundary of the efficiency map. The limiting characteristic curve in the low speed range is limited by the maximum available current of the utilized inverter , while in the high speed range or rather field weakening area it is determined by both the maximum voltage and the maximum current of the inverter.

The characteristic curves are illustrated in Fig. 9.28. It can be seen that the difference of the torque at low speed range is relatively small, which equals 5.4 Nm for the lowest speed. Furthermore, it can be noted the characteristic speed for both simulation and measurement is nearly the same equal to $2600\,\mathrm{min}^{-1}$.

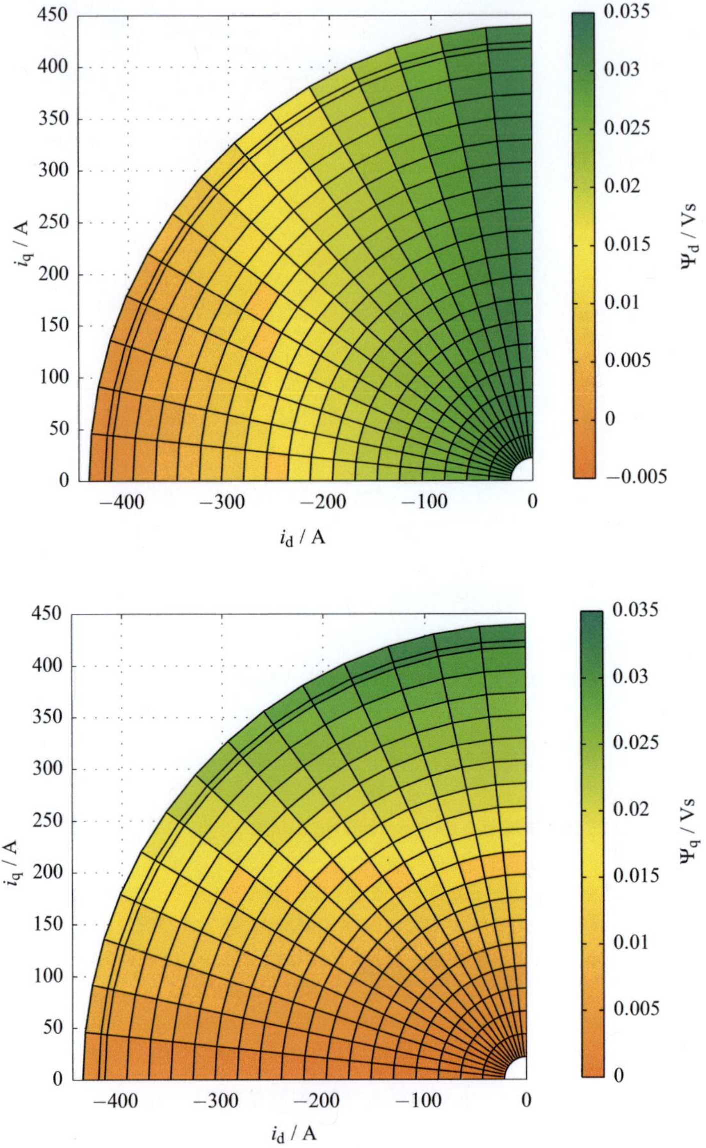

Figure 9.27. Flux linkage Ψ_d and Ψ_q calculated with 3D FEM

When the rotational speed of the prototype becomes higher, the deviation of the torque increases significantly. The most important reason is that the controlling quality of current becomes poorer and there are more harmonics contained in the current, which result in more eddy current losses in PM and higher loss of torque. Above the characteristic speed, the output torque becomes smaller because the total output torque becomes smaller due to the increasing current component i_d. The deviation between the simulation and measurement remains almost unchanged equal to 20 Nm

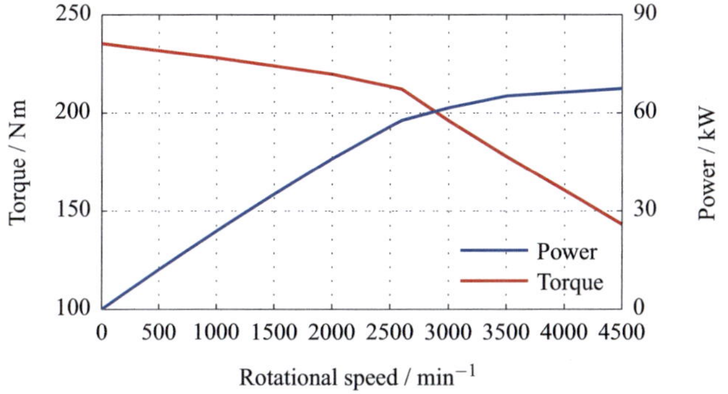

(a) Measurement results

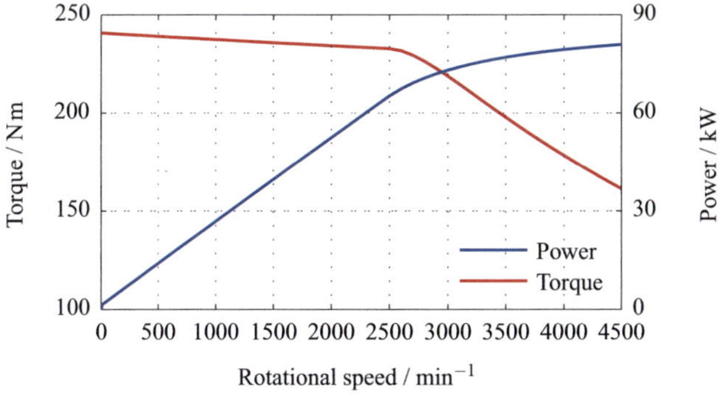

(b) Simulation results

Figure 9.28. Comparison of limiting characteristic curves

In similar, the deviation of power becomes higher with increasing rotational speed. It equals 9.65 kW at the characteristic speed and reaches the maximum equal to 13.47 kW at 4500 min^{-1}.

Measurement of efficiency map

The efficiency map is the most informative property of an electrical machine, which is determined based on the measurement of all operating points of the prototype at different rotational speed. The measured efficiency maps is illustrated in Fig. 9.29.

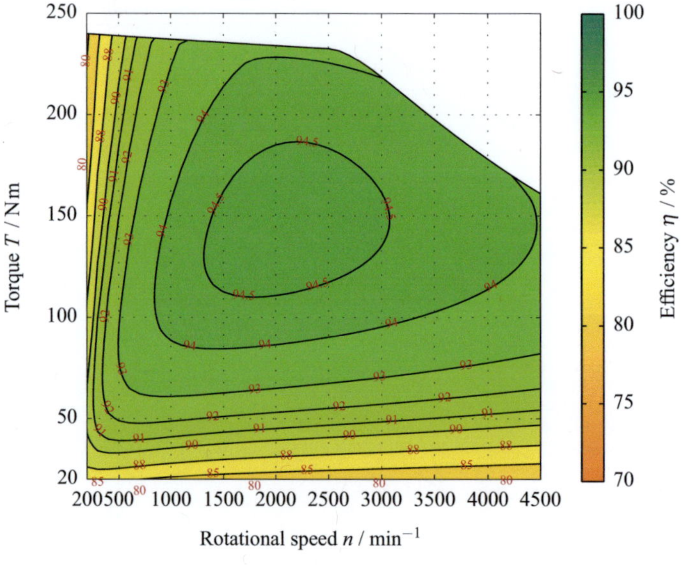

Figure 9.29. Measured efficiency map of the prototype

In comparison, the efficiency map based on the 3D FEM is illustrated in Fig. 9.30. It can be seen that the efficiency becomes lower with increasing speed and higher current. There is a deviation of peak efficiency equal to 2 % between simulation and measurement. The most important reasons are listed as follows:

- Inadequate controlling quality of current, which leads to many harmonics in the current. Therefore, the useful harmonic generating the torque is smaller.

- The harmonics contained in the current cause higher copper losses, higher eddy current losses in PM and higher iron losses in SMC.
- With increasing rotational speed, the friction losses caused by the bearings and the air flow become higher as well.
- Due to higher temperature of prototype, properties of PM deteriorate as well.

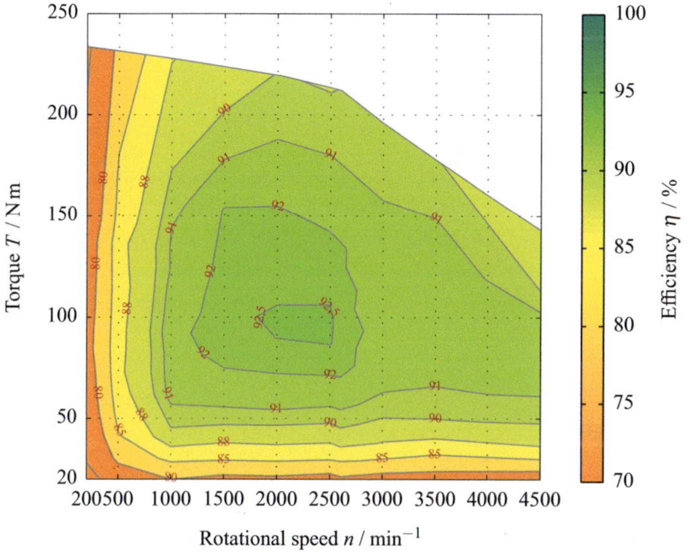

Figure 9.30. Simulated efficiency map of the prototype

Based on the above results, it is clear that a segmentation of PM is necessary for the PMAFM-ISA to reduce the induced eddy currents. On one hand, there are more harmonics contained in the MMF of concentrated winding compared with distributed winding. On the other hand, the phase current isn't ideally sinusoidal as in FEM and contains a lot of harmonics in the measurement.

In addition, the current control of the PMAFM-ISA should be further improved to achieve a phase current with less harmonics. At last, it should be noted that the efficiency of prototype can be further improved, if the SMC components aren't manufactured from the cylinder but directly pressed.

Temperature measurement of prototype with electrical loading

The Temperature of the prototype with different electrical loading at various speed is another important aspect to evaluate its performance, which can determine the continuous output power. Afterwards, the results based on the coupled thermal and fluid dynamic analysis are compared with the measurements.

Similar to the measurement of prototype at no-load, the position 1 corresponds to the position in the inner radius of the stator, while position 2 is defined as the position between the stator segments at the average radius of the stator. Furthermore, the thermal source is defined as the actual measured power losses instead of the simulation results, in order to eliminate the errors resulted by the deviation of power losses.

Operating point	**Numerical**		**Measurement**	
	Position1	Position2	Position1	Position2
$200\,\mathrm{min}^{-1}$ $i_\mathrm{q} = 150\,\mathrm{A}$	45.16 °C	43.32 °C	49.81 °C	48.21 °C
$200\,\mathrm{min}^{-1}$ $i_\mathrm{q} = 200\,\mathrm{A}$	59.95 °C	57.10 °C	67.90 °C	64.57 °C
$200\,\mathrm{min}^{-1}$ $i_\mathrm{q} = 250\,\mathrm{A}$	82.21 °C	77.77 °C	96.24 °C	89.47 °C
$200\,\mathrm{min}^{-1}$ $i_\mathrm{q} = 300\,\mathrm{A}$	115.08 °C	107.68 °C	130.1 °C	121.5 °C
$500\,\mathrm{min}^{-1}$ $i_\mathrm{q} = 300\,\mathrm{A}$	127.37 °C	119.13 °C	148.0 °C	139.1 °C
$1000\,\mathrm{min}^{-1}$ $i_\mathrm{q} = 300\,\mathrm{A}$	165.35 °C	155.29 °C	165.2 °C	157.4 °C

Table 9.3. Comparison of the temperature of prototype with $i_\mathrm{d} = 0\,\mathrm{A}$

It can be noted that the deviation becomes larger compared with the measurement without electrical loading, especially at low speed and high electrical loading. The maximum deviation equal to 20 °C occurs at the low speed $500\,\mathrm{min}^{-1}$. In addition to the reasons mentioned above, another important reason is that the accurate distribution of the measured total power losses is hard to determine, which has however obvious influence on the FEM results.

10

Conclusions and Future Work

In this chapter, the performed research work and the corresponding results are briefly summarised. During the prototyping and measurement of the prototype, some aspects that can improve the performance of the PMAFM-ISA have been noted. At last, the improvements that may be helpful for the future work are described.

10.1 Conclusions

The application of SMC in the electrical traction machine for hybrid electrical vehicle or electrical vehicle has been researched in this doctoral thesis. The fundamental properties of SMC were measured with different ring specimens. On this basis, two electrical machines, namely a permanent magnet excited axial flux machine with internal segmented armature (PMAFM-ISA) and a permanent magnet excited transverse flux machine with claw poles (PMTFM-CP) have been designed. They were compared with the reference machine, namely the current design of PMRFM of industrial partner. Finally, a prototype of the PMAFM-ISA topology has been manufactured and measured. Based on the performed work, the most important conclusions are presented as follows.

Through the comparison of the measured electromagnetic properties of SMC with electrical steel, it is clear that the unsaturated permeability of SMC is lower than electrical steel due to the insulation layer around the iron particles. Besides, the hysteresis losses of SMC are also higher which are the dominant components of the iron losses in the low frequency domain. However, the SMC demonstrates higher electrical resistance and lower eddy current losses. For instance, the iron losses of SMC are almost the same as that of the electrical steel M330-35A when the frequency and flux density equals 1 kHz and 1.5 T. It should also be noted that for the electrical steel NO20 with the thickness equal to 0.20 mm, the iron losses remain lower than SMC, even at extremely high frequency.

In order to calculate the iron losses of SMC parts in electrical machines, an iron loss model based on the Bertotti formula and FEM has been developed. A test system comprising a yoke and a tooth made from the same SMC has been constructed and built. Through the comparison between the measured and the calculated iron losses based on the developed model, it can be noted the deviation is lower than 8 % for the frequencies lower than 500 Hz. However, the deviation increases with higher frequency due to the machining of SMC parts, the inaccuracy of current measurement, the existence of rotational field and so on. In spite of the above errors, the measurement matches relatively well with the FEA, which implies that the developed iron loss model is valid and can be used to calculate the iron losses of electrical machines.

For the purpose of efficient analysis of the novel electrical machines, a simulation platform based on commercially available 3D FEM software has been developed. With its help, the script-based modelling, solving and post-processing are enabled. Considering the 3D magnetic flux, time-consuming 3D FEM is required to evaluate the performance of both PMAFM and PMTFM accurately. In order to reduce the computation time of PMAFM, a method for the iron losses calculation of PMAFM with different winding layouts has been developed and validated with FEM. With its help, the time needed to analyse a PMAFM can be reduced to $1/6$.

With thc help of the simulation platform, a PMAFM-ISA and a PMTFM-CP have been designed and compared with the current design of PMRFM. Based on the detailed comparison, it can be concluded that PMAFM with surface mounted PMs shows higher cogging torque and no-load losses than the PMTFM and PMRFM with embedded PMs. Compared with PMRFM, the PMAFM possesses a better limiting characteristic curve especially in the field weakening area, which implies a higher peak torque and output power. The PMTFM with higher pole pair number can offer higher torque density than the PMAFM-ISA and PMRFM for the same current density. However, the output power of PMTFM is much lower due to its low power factor. High efficiency can be achieved for all researched topologies. The PMTFM has the highest peak efficiency. This is because the PMs are embedded in the rotor, and the ring winding results in low copper losses. The efficiency of PMAFM-ISA can be effectively improved through the segmentation of PM.

In order to validate the FEM results, a mechanical construction has been developed for the topology PMAFM-ISA and validated in terms of the mechanical stress and deformation. Because the properties of both PM and adhesive depend strongly on the temperature, a thermal model based on the coupled thermal and fluid dynamic analysis and an analytical thermal model have been developed. With their help, it is noted that the stationary temperatures of both PM and stator segments reduce significantly by means of PM segmentation, meaning that the continuous power of PMAFM-ISA can be effectively improved.

Based on the developed mechanical construction, a prototype has been manufactured and measured with the test stand developed at the institute of electrical engineering (ETI). The measurements match well with the simulations at no-load condition. When the prototype is fed with inverter, the deviation becomes larger with increasing speed and higher electrical loading. The most important reason is that many harmonics are contained in the phase current due to the pulse width modulation of the inverter, especially at the high speed considering its high pole pair number. As a result, more PM eddy current losses and iron losses in SMC are generated, which deteriorates the PMAFM-ISA performance obviously. For instance, reduction of peak efficiency by 2 % can be observed. The developed thermal model is also validated by measurement.

In summary, the segmentation of PM is of great importance for the PMAFM-ISA with surface mounted PMs and concentrated winding based on the electromagnetic analysis, the thermal analysis and the measurement results. However, it should be noted that the mechanical firmness of PM becomes worse. Thus, more attentions need to be paid to the mechanical analysis.

Based on above analysis, it can be concluded that for the specific applications with special geometrical constraints or requirements, some novel electrical machines such as PMAFM and PMTFM can be competitive compared with the traditional radial flux machine. In theses novel machines, the application of SMC becomes attractive or even necessary considering the existence of complex components and 3D magnetic flux. Furthermore, great attentions should be paid to the electromagnetic design and mechanical construction to utilize the special advantages of SMC.

10.2 Future work

The performed research work has demonstrated the potential of SMC to be used in the novel electrical machines with 3D magnetic flux. During the electromagnetic design, the mechanical construction and analysis, and the measurement of the first prototype manufactured in ETI, a lot of valuable experiences have been gained. The aspects to improve the performance of PMAFM-ISA are summarized, which can be classified into optimization of electromagnetic design and optimization of mechanical construction [S7].

Optimization of electromagnetic design

The current electromagnetic design of PMAFM-ISA can be optimized considering the rotor yoke material, the DLCW with different layout and the segmentation of PM.

- Rotor yoke material

  In the current electromagnetic design, the rotor yoke consists of 12 segments due to its large size. For this reason, a rotor carrier made from aluminium is needed to fix the rotor yoke segments, which leads to a complex mechanical construction and higher thermal resistance. Based on above consideration, it is considerable to utilize the electrical steel illustrated in Fig. 10.1 to manufacture the rotor yoke.

  First of all, it brings in an improvement of the mechanical stiffness. On the other hand, the iron losses of the rotor yoke would reduce as well. Of course, it is necessary to perform the similar adhesive test to examine whether the PM can be firmly glued on the rotor yoke at high temperature.

- DLCW with elementary winding 2B2

  During the measurement, it is noted that a higher pole pair number of PMAFM-ISA leads to a more difficult current controlling of the inverter due to the limited sampling frequency. Therefore, it would be helpful to reduce the pole pair number of the PMAFM-ISA. In addition, another disadvantage of the DLCW with the elementary winding 2A1 is the high torque ripple. In comparison, the DLCW with the elementary winding 2B2 can offer a much lower torque ripple, and the skewing of the PM is thus not necessary any more.

(a) Unfixed ring (b) Fixed ring

Figure 10.1. Rotor yoke made from electrical steel band

- Segmentation of PM

  Based on the measurements, it is noted that a segmentation of PM is necessary to reduce the induced eddy currents, especially for the PMAFM-ISA with high pole pair number and concentrated winding. As mentioned above, more attentions should be paid to the mechanical analysis due to the deterioration of the mechanical property of PM.

Optimization of mechanical design

From the measurement of the prototype, it can be clearly seen that the temperature is critical for the performance and functionality of the prototype. For this reason, the current mechanical construction should be improved to achieve better air cooling effect and to reduce the costs.

- POM instead of PEEK

  In the current mechanical design, four toroidal plates made from the material polyetheretherketone (PEEK) is used to fix the SMC stator segments and the PMs. However, the PEEK plate is very expensive and difficult to manufacture. For this reason, it is important to find another material to replace the PEEK.

  The Polyoxymethylene (POM) illustrated in Fig. 10.2 is an outstanding candidate because the material is much cheaper and easier to be machined.

Another great advantage is that the material can be easily separated from the cast stator, which results in a creation of additional cooling channels in the air gap.

- Bearing system

  The accurate spindle bearings are currently utilized as the fixed bearing to make the air gaps as symmetrical as possible. However, this kind of bearing is expensive and demands high manufacturing accuracy of the contacting surfaces. For this reason, the double row angular contact ball bearing illustrated in Fig. 10.2 seems to be a feasible alternative considering its axial and radial load capacity. However, its deformation and lifetime should be carefully examined before being employed in the prototype.

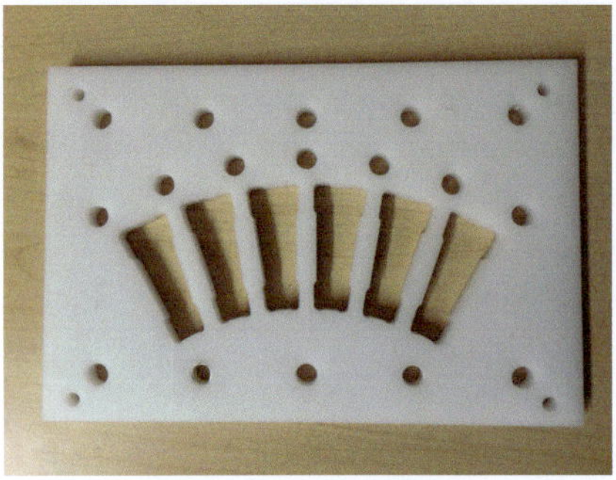

(a) Plate made from POM

(b) Double row angular contact ball bearing

Figure 10.2. Alternative material and bearings

Based on above considerations, a PMAFM-ISA with 10 pole pairs and DLCW $2 \times 2B2$ has been designed. It has two rotor yokes made from electrical steel, and the PMs aren't skewed but radially segmented into 8 parts. The corresponding 3D FEM model, the mechanical construction and the efficiency map are illustrated in Fig. 10.3. The efficiency map reveals great optimization potential of the first prototype.

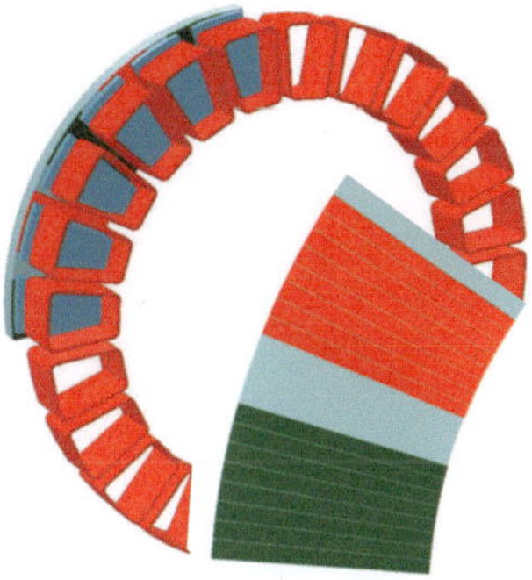

(a) Electromagnwetic design

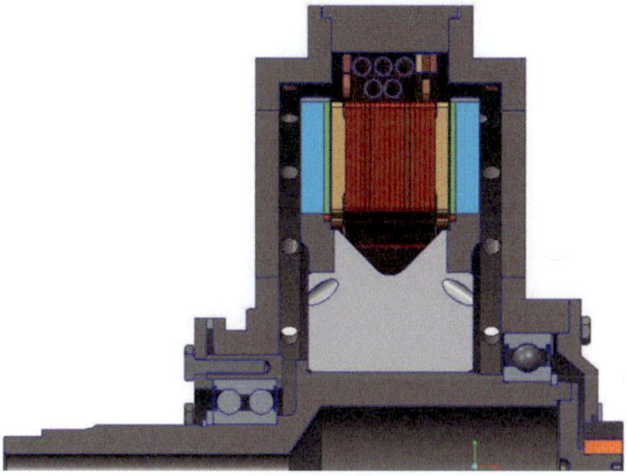

(b) Mechanical construction

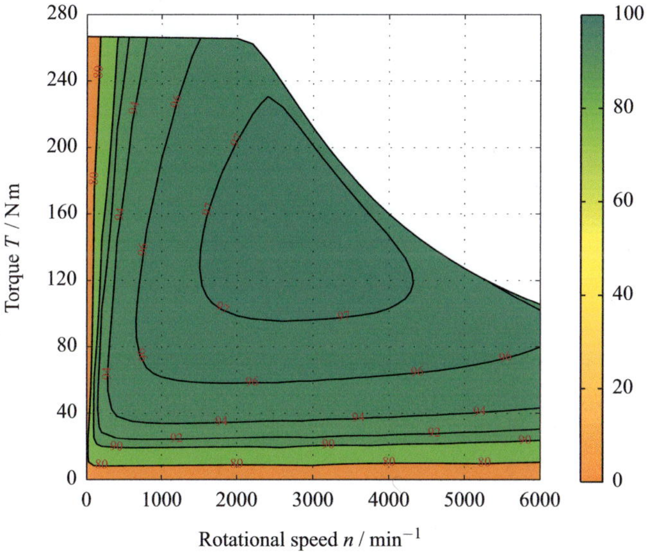

(c) Efficiency map

Figure 10.3. Optimized design of PMAFM-ISA

List of Figures

List of Tables

Bibliography

Own publications within the doctoral thesis

[O1] B. ZHANG and M. DOPPELBAUER, "Development of a simulation platform for modeling and analysis of transverse flux machine based on 3d fem", in *International ETG congress (Berlin)*, Nov. 2013.

[O2] B. ZHANG, J. LIU, and M. DOPPELBAUER, "A comparison between stationary and transient analysis of electrical machines using 3d fea", in *Electrical Machines (ICEM), 2014 International Conference on*, Sep. 2014.

[O3] B. ZHANG and M. DOPPELBAUER, "Numerical iron loss calculation of a new axial flux machine with segmented-armature-torus topology", in *Power Electronics, Machines and Drives (PEMD 2014), the 7th IET International Conference on*, Apr. 2014.

[O4] B. ZHANG, T. TALMON, and M. DOPPELBAUER, "Iron losses calculation of a novel axial flux machine with different slot-pole combinations based on the fea corresponding to one-sixth electrical period", in *Power Electronics, Machines and Drives (PEMD 2016), the 8th IET International Conference on*, Apr. 2016.

[O5] B. ZHANG and M. DOPPELBAUER, "Iron losses calculation of an axial flux machine based on 3d fea results corresponding to one-sixth electrical period", *IEEE TRANSACTIONS ON ENERGY CONVERSION*, 2017, to be published.

[O6] B. ZHANG, M. KAHLE, and M. DOPPELBAUER, "Design, optimization and comparison of transverse flux machine with surface mounted magnets and flux concentration", in *VDE/VDI-Conference: Drive Systems (Nürtingen)*, Sep. 2013.

[O7] B. ZHANG, T. EPSKAMP, M. DOPPELBAUER, and M. GREGOR, "A comparison of the transverse, axial and radial flux pm synchronous motors for electric vehicle", in *Electric Vehicle Conference (IEVC), 2014 IEEE International*, Dec. 2014.

[O8] B. ZHANG, A. WANG, and M. DOPPELBAUER, "Multi-objective optimization of a novel transverse flux machine with claw pole and flux-concentrating structure", *IEEE TRANSACTIONS ON MAGNETICS*, vol. 52, pp. 1–10, 8 2016, to be published.

[O9] B. ZHANG, A. S. WANG, and M. DOPPELBAUER, "Optimization of a transverse flux machine with claw pole and flux-concentrating structure", in *Electric Machines & Drives Conference (IEMDC), 2015 IEEE International*, May 2015.

[O10] B. ZHANG, T. SEIDLER, R. DIERKEN, and M. DOPPELBAUER, "Development of a yokeless and segmented armature axial flux machine", *IEEE TRANSACTIONS ON INDUSTRIAL ELECTRONICS*, vol. 63, pp. 2062–2071, 4 2016.

[O11] B. ZHANG, Y. WANG, and M. DOPPELBAUER, "Mechanical construction and analysis of an axial flux segmented armature torus machine", in *Electrical Machines (ICEM), 2014 International Conference on*, Sep. 2014.

[O12] B. ZHANG, M. DOPPELBAUER, A. SCHOPPA, and M. GREGOR, "Development of a novel yokeless and segmented armature axial flux machine based on soft magnetic composites", in *World PM2016 Congress & Exhibition*, to be published, Oct. 2016.

Supervised students theses

[S1] M. VIVEKANANTHAN, "Investigation of the machining influence on the electromagnetic properties of components made from soft magnetic composites", Bachelor thesis, Karlsruhe Institute of Technology, Institute of Electrical Engineering, 2015.

[S2] J. LIU, "Further development of a simulation platform and mechanical construction of an axial flux machine with inner rotor", Master thesis, Karlsruhe Institute of Technology, Institute of Electrical Engineering, 2014.

[S3] T. TALMON, "Investigation of the double layer concentrated winding for the application in the axial flux machine based on smc", Diploma thesis, Karlsruhe Institute of Technology, Institute of Electrical Engineering, 2014.

[S4] H. JIANG, "Design of an axial flux machine based on the soft magnetic composites", Diploma thesis, Karlsruhe Institute of Technology, Institute of Electrical Engineering, 2012.

[S5] S. KÖRNER, "Analysis and comparison of axial and radial flux machine", Bachelor thesis, Karlsruhe Institute of Technology, Institute of Electrical Engineering, 2015.

[S6] Q. ZHANG, "Investigation of the application of 2d fem in the analysis of axial flux machine with segmented-armature-torus structure", Bachelor thesis, Karlsruhe Institute of Technology, Institute of Electrical Engineering, 2015.

[S7] J. BÄR, "Comprehensive optimization of an axial flow machine with segmented-armature-torus topology", Master thesis, Karlsruhe Institute of Technology, Institute of Electrical Engineering, 2016.

[S8] M. KAHLE, "Design, optimization and comparison of transverse flux machine with surface mounted pm and flux concentrating structure", Bachelor thesis, Karlsruhe Institute of Technology, Institute of Electrical Engineering, 2013.

[S9] J. BÄR, "Design and optimization of a transverse flux machine with flux concentrating and claw pole structure", Bachelor thesis, Karlsruhe Institute of Technology, Institute of Electrical Engineering, 2013.

[S10] L. SUN, "Modelling and comparison of three fundamental topologies of transverse flux machine with fem", Bachelor thesis, Karlsruhe Institute of Technology, Institute of Electrical Engineering, 2012.

[S11] A. WANG, “Optimization of a claw-pole transverse flux machine with flux concentrating structure”, Diploma thesis, Karlsruhe Institute of Technology, Institute of Electrical Engineering, 2014.

[S12] Y. WANG, “Mechanical construction and analysis of an axial flux machine with segmented-armature-torus topology”, Master thesis, Karlsruhe Institute of Technology, Institute of Electrical Engineering, 2014.

[S13] R. DIERKEN, “Analysis and prototyping of an axial flux machine with segmented-armature-torus structure”, Bachelor thesis, Karlsruhe Institute of Technology, Institute of Electrical Engineering, 2014.

[S14] S. ZAKARIA, “Coupling of aerothermal simulation with electromagnetic design of electrical machines”, Diploma thesis, Karlsruhe Institute of Technology, Institute of Electrical Engineering, 2014.

[S15] T. SEIDLER, “Thermal analysis of an axial flux machine with segmented-armature-torus topology”, Bachelor thesis, Karlsruhe Institute of Technology, Institute of Electrical Engineering, 2014.

Other publications

[1] C. CHAN and K. CHAU, *Modern electric vehicle technology*. Oxford University Press, 2001, vol. 47.

[2] K.-T. CHAU, W. LI, and C. H. LEE, "Challenges and opportunities of electric machines for renewable energy", *Progress In Electromagnetics Research B*, vol. 42, pp. 45–74, 2012.

[3] M. COSSALE, A. KRINGS, J. SOULARD, A. BOGLIETTI, and A. CAVAGNINO, "Practical investigations on cobalt-iron laminations for electrical machines", in *Electrical Machines (ICEM), 2014 International Conference on*, Sep. 2014, pp. 1390–1395.

[4] M. LINDNER, P. BRAUER, and R. WERNER, "Increasing the torque density of permanent-magnet synchronous machines using innovative materials and winding technologies", in *Systems, Signals and Devices (SSD), 2012 9th International Multi-Conference on*, Mar. 2012, pp. 1–7.

[5] A. BOEHM and I. HAHN, "Comparison of soft magnetic composites (smcs) and electrical steel", in *Electric Drives Production Conference (EDPC), 2012 2nd International*, Oct. 2012, pp. 1–6.

[6] W. PLUTA, "Some properties of factors of specific total loss components in electrical steel", *Magnetics, IEEE Transactions on*, vol. 46, pp. 322–325, Feb. 2010.

[7] Z. WANG, Y. ENOMOTO, M. ITO, R. MASAKI, S. MORINAGA, H. ITABASHI, and S. TANIGAWA, "Development of an axial gap motor with amorphous metal cores", in *Electrical Machines and Systems, 2009. ICEMS 2009. International Conference on*, Nov. 2009, pp. 1–6.

[8] H. SHOKROLLAHI and K. JANGHORBAN, "Soft magnetic composite materials (smcs)", *Journal of Materials Processing Technology*, vol. 189, no. 1, pp. 1–12, 2007.

[9] O. ANDERSSON and P. HOFECKER, "Advances in soft magnetic composites - materials and applications", *PowderMet2009*, pp. 1–12, Jul. 2009.

[10] L. HULTMAN and H. LOPES, "Existing and future automotive applications for soft magnetic composites", *Powder Metallurgy (PM)*, 2008.

[11] A. SCHOPPA and P. DELARBRE, "Soft magnetic powder composites and potential applications in modern electric machines and devices", *Magnetics, IEEE Transactions on*, vol. 50, pp. 1–4, Apr. 2014.

[12] L. PENNANDER and A. JACK, "Soft magnetic iron powder materials ac properties and their application in electrical machines", *Magnetic Materials*,

[13] J. WANG, X. YUAN, and K. ATALLAH, "Design optimization of a surface-mounted permanent-magnet motor with concentrated windings for electric vehicle applications", *Vehicular Technology, IEEE Transactions on*, vol. 62, pp. 1053–1064, Mar. 2013.

[14] S. HUANG, J. LUO, F. LEONARDI, and T. LIPO, "A general approach to sizing and power density equations for comparison of electrical machines", *Industry Applications, IEEE Transactions on*, vol. 34, pp. 92–97, Jan. 1998, ISSN: 0093-9994.

[15] D. DORRELL, A. KNIGHT, L. EVANS, and M. POPESCU, "Analysis and design techniques applied to hybrid vehicle drive machines-assessment of alternative ipm and induction motor topologies", *Industrial Electronics, IEEE Transactions on*, vol. 59, pp. 3690–3699, Oct. 2012, ISSN: 0278-0046.

[16] T. GEMASSMER, "Effiziente und dynamische drehmomenteinprägung in hoch ausgenutzten synchronmaschinen mit eingebetteten magneten", PhD thesis, Karlsruher Instituts fÃ¼r Technologie, 2015.

[17] Z. ZHU and C. CHAN, "Electrical machine topologies and technologies for electric, hybrid, and fuel cell vehicles", in *Vehicle Power and Propulsion Conference, 2008. VPPC '08. IEEE*, Sep. 2008, pp. 1–6.

[18] F. MARQUEZ, "Electric traction machine design for an e-rwd unit", PhD thesis, 2014.

[19] F. PROFUMO, Z. ZHANG, and A. TENCONI, "Axial flux machines drives: A new viable solution for electric cars", *IEEE TRANSACTIONS ON INDUSTRIAL ELECTRONICS*, vol. 44, pp. 39–45, 1997.

[20] W. FEI, P. LUK, and K. JINUPUN, "A new axial flux permanent magnet segmented-armature-torus machine for in-wheel direct drive applications", in *Power Electronics Specialists Conference, 2008. PESC 2008. IEEE*, Jun. 2008, pp. 2197–2202.

[21] Y. KOKUBO and S. SHIMOMURA, "Design of dual rotor-axial gap pmvm for hybrid electric vehicle", in *Electrical Machines and Systems (ICEMS), 2014 17th International Conference on*, Oct. 2014, pp. 2573–2578.

[22] W. ZHAO, A. AHMED, I. HUSAIN, and E. MULJADI, "A novel transverse flux machine for vehicle traction aplications", in *Power Energy Society General Meeting, 2015 IEEE*, Jul. 2015, pp. 1–5.

[23] P. SEIBOLD, F. SCHULLER, M. BEEZ, and N. PARSPOUR, “Design and measurement of a laminated permanent magnet excited transverse flux machine for electrical vehicles”, in *Electric Drives Production Conference (EDPC), 2014 4th International*, Sep. 2014, pp. 1–6.

[24] P. SEIBOLD, F. SCHULLER, N. PARSPOUR, and M. GÄRTNER, “Transverse flux machine in a diesel-electric drive system”, *ATZoffhighway*, vol. 5, no. 2, pp. 68–77, 2012.

[25] P. SEIBOLD, M. GARTNER, F. SCHULLER, and N. PARSPOUR, “Design of a transverse flux permanent magnet excited machine as a near-wheel motor for the use in electric vehicles”, in *Electrical Machines (ICEM), 2012 XXth International Conference on*, Sep. 2012, pp. 2641–2646.

[26] D. PLATT and B. SMITH, “Twin rotor drive for an electric vehicle”, *Electric Power Applications, IEE Proceedings B*, vol. 140, no. 2, pp. 131–138, Mar. 1993.

[27] C. LEE, K. CHAU, C. LIU, T. CHING, and F. LI, “A high-torque magnetless axial-flux doubly salient machine for in-wheel direct drive applications”, *Magnetics, IEEE Transactions on*, vol. 50, no. 11, pp. 1–5, Nov. 2014.

[28] R. MADHAVAN and B. FERNANDES, “A novel axial flux segmented srm for electric vehicle application”, in *Electrical Machines (ICEM), 2010 XIX International Conference on*, Sep. 2010, pp. 1–6.

[29] T. LAMBERT, M. BIGLARBEGIAN, and S. MAHMUD, “A novel approach to the design of axial-flux switched-reluctance motors”, *Machines*, vol. 3, no. 1, pp. 27–54, 2015.

[30] R. MADHAVAN and B. FERNANDES, “Comparative analysis of axial flux srm topologies for electric vehicle application”, in *Power Electronics, Drives and Energy Systems (PEDES), 2012 IEEE International Conference on*, Dec. 2012, pp. 1–6.

[31] Y. GUO, J. ZHU, D. DORRELL, H. Y. LU, and Y. WANG, “Development of a claw pole permanent magnet motor with a molded low-density soft magnetic composite stator core”, in *Energy Conversion Congress and Exposition, 2009. ECCE 2009. IEEE*, Sep. 2009, pp. 294–301.

[32] Y. GUO, J. ZHU, and D. DORRELL, “Design and analysis of a claw pole permanent magnet motor with molded soft magnetic composite core”, *Magnetics, IEEE Transactions on*, vol. 45, no. 10, pp. 4582–4585, Oct. 2009.

[33] M. PERSSON, P. JANSSON, A. JACK, and B. MECROW, “Soft magnetic composite materials-use for electrical machines”, in *Electrical Ma-*

chines and Drives, 1995. Seventh International Conference on (Conf. Publ. No. 412), Sep. 1995, pp. 242–246.

[34] M. PERSSON and P. JANSSON, "Advances in powder metallurgy soft magnetic composite materials for electrical machines", in *Impact of New Materials on Design, IEE Colloquium on*, vol. 4, Dec. 1995, pp. 1–6.

[35] J. G. ZHU, Y. G. GUO, Z. W. LIN, Y. J. LI, and Y. K. HUANG, "Development of pm transverse flux motors with soft magnetic composite cores", *Magnetics, IEEE Transactions on*, vol. 47, no. 10, pp. 4376–4383, Oct. 2011.

[36] M. LEVIN, "Charakterisierung von weichmagnetischen pulververbundwerkstoffen für den einsatz in traktionsantrieben", PhD thesis, Technische Universität München, 2012.

[37] S. SATO, S. TAKEGUCHI, A. MATSUO, and T. ISHIKAWA, "3-d soft magnetic composite core superior in the downsizing of motors", in *Proc. 20th MAGDA Conf. Pacific Asia, Taipei*, Nov. 2011, pp. 355–360.

[38] T. ISHIKAWA, S. SATO, S. TAKEGUCHI, and A. MATSUO, "Design of a dc motor made of soft magnetic composite core by the experimental design method", *Magnetics, IEEE Transactions on*, vol. 48, pp. 3132–3135, Nov. 2012.

[39] J. CROS, P. VIAROUGE, and A. HALILA, "Brush dc motors with concentrated windings and soft magnetic composites armatures", in *Industry Applications Conference, 2001. Thirty-Sixth IAS Annual Meeting. Conference Record of the 2001 IEEE*, vol. 4, Sep. 2001, pp. 2549–2556.

[40] A. JACK, B. MECROW, C. MADDISON, and N. WAHAB, "Claw pole armature permanent magnet machines exploiting soft iron powder metallurgy", in *Electric Machines and Drives Conference Record, 1997. IEEE International*, May 1997, MA1/5.1–MA1/5.3.

[41] A. JACK, B. MECROW, P. DICKINSON, P. JANSSON, and L. HULTMAN, "Design and testing of a universal motor using a soft magnetic composite stator", in *Industry Applications Conference, 2000. Conference Record of the 2000 IEEE*, vol. 1, 2000, pp. 46–50.

[42] R. WANG and R. WALTER, "Modeling of universal motor performance and brush commutation using finite element computed inductance and resistance matrices", *Energy Conversion, IEEE Transactions on*, vol. 15, no. 3, pp. 257–263, Sep. 2000.

[43] Ö. KROGEN and A. G. JACK, "Insulated iron powder (smc) used as soft magnetic material in a rotating electrical machine", *Proc. Metallurgy World*, 2000.

[44] J. CROS, P. VIAROUGE, Y. CHALIFOUR, and J. FIGUEROA, "A new structure of universal motor using soft magnetic composites", *Industry Applications, IEEE Transactions on*, vol. 40, no. 2, pp. 550–557, Mar. 2004.

[45] A. JACK, "Experience with using soft magnetic composites for electrical machines", in *New Magnetic Materials - Bonded Iron, Lamination Steels, Sintered Iron and Permanent Magnets (Digest NMo. 1998/259), IEE Colloquium on*, May 1998, pp. 3/1–3/4.

[46] C. LEE, K. CHAU, and C. H. LIU, "Design and analysis of a dc field multitooth switched reluctance machine by using soft-magnetic-composite material", in *Industrial Electronics (ISIE), 2013 IEEE International Symposium on*, May 2013, pp. 1–6.

[47] K. VIJAYAKUMAR, R. KARTHIKEYAN, G. SATHISHKUMAR, and R. ARUMUGAM, "Torque-to-weight ratio improvement with soft magnetic composite material in high speed switched reluctance motor", in *TENCON 2008 - 2008 IEEE Region 10 Conference*, Nov. 2008, pp. 1–4.

[48] K. VIJAYAKUMAR, R. KARTHIKEYAN, and R. ARUMUGAM, "Influence of soft magnetic composite material on the electromagnetic torque characteristics of switched reluctance motor", in *Power System Technology and IEEE Power India Conference, 2008. POWERCON 2008. Joint International Conference on*, Oct. 2008, pp. 1–6.

[49] K. VIJAYAKUMAR, R. KARTHIKEYAN, G. SATHISHKUMAR, and R. ARUMUGAM, "Two dimensional magnetic and thermal analysis of high speed switched reluctance motor using soft magnetic composite material", in *TENCON 2008 - 2008 IEEE Region 10 Conference*, Nov. 2008, pp. 1–5.

[50] J. REITER F. and T. STUART, "Composite powder metal synchronous reluctance machine", in *Electric Machines and Drives Conference, 2003. IEMDC'03. IEEE International*, vol. 1, Jun. 2003, pp. 523–527.

[51] K. VIJAYAKUMAR, R. KARTHIKEYAN, S. KANNAN, G. SUNDER, and R. ARUMUGAM, "Dynamic analysis of switched reluctance motor using soft magnetic composite material", in *Power Electronics, Drives and Energy Systems (PEDES) 2010 Power India, 2010 Joint International Conference on*, Dec. 2010, pp. 1–6.

[52] M. POPESCU, T. JOKINEN, E. DEMETER, and V. NAVRAPESCU, "Magnetic materials in design and construction of electrical motors", in *Electrotechnical Conference, 1998. MELECON 98., 9th Mediterranean*, vol. 1, May 1998, pp. 125–128.

[53] T. FUKUDA and M. MORIMOTO, "Load characteristics of induction motor made of soft magnetic composite (smc)", in *Electrical Machines and Systems, 2008. ICEMS 2008. International Conference on*, Oct. 2008, pp. 53–56.

[54] M. MORIMOTO, "Induction motor made of iron powder core", in *Power Electronics Conference (IPEC), 2010 International*, Jun. 2010, pp. 1814–1817.

[55] T. MATSUMOTO, S. SHIMOMURA, and M. MORIMOTO, "Study of iron loss in induction motor made with soft magnetic composite core based on finite element analysis", in *Power Electronics Conference (IPEC), 2010 International*, Jun. 2010, pp. 270–277.

[56] Y. SASAKI and M. MORIMOTO, "The inverter drive loss of induction motor made by soft magnetic composite", in *Power Electronics and Drive Systems, 2009. PEDS 2009. International Conference on*, Nov. 2009, pp. 1425–1428.

[57] M. MORIMOTO, "Performance improvement of induction motor made of three dimensional shaped smc core", in *Power Electronics and Drive Systems (PEDS), 2011 IEEE Ninth International Conference on*, Dec. 2011, pp. 162–166.

[58] N. TSUCHIYA and M. MORIMOTO, "Iron loss characteristics of induction motor made of soft magnetic composite (smc)", in *Electrical Machines and Systems (ICEMS), 2010 International Conference on*, Oct. 2010, pp. 1337–1340.

[59] M. MORIMOTO, "Efficiency improvement of induction motor by 3-d core made of smc", in *Power Electronics and Motion Control Conference (EPE/PEMC), 2012 15th International*, Sep. 2012, LS1b.3–1–LS1b.3–5.

[60] T. IWAWAKI, M. INAMORI, and M. MORIMOTO, "Variation of iron loss by manufacturing method of smc motor core", in *Electrical Machines and Systems (ICEMS), 2013 International Conference on*, Oct. 2013, pp. 2040–2043.

[61] J. CROS, P. VIAROUGE, and C. GÉLINAS, "Design of pm brushless motors using iron-resin composites for automotive applications", in *Industry Applications Conference, 1998. Thirty-Third IAS Annual Meeting. The 1998 IEEE*, vol. 1, Oct. 1998, pp. 5–11.

[62] J. CROS and P. VIAROUGE, "New structures of polyphase claw-pole machines", *Industry Applications, IEEE Transactions on*, vol. 40, no. 1, pp. 113–120, Jan. 2004.

[63] Y. Enomoto, M. Ito, H. Koharagi, R. Masaki, S. Ohiwa, C. Ishihara, and M. Mita, "Evaluation of experimental permanent-magnet brushless motor utilizing new magnetic material for stator core teeth", *Magnetics, IEEE Transactions on*, vol. 41, no. 11, pp. 4304–4308, Nov. 2005.

[64] G. Liew, E. Tsang, N. Ertugrul, W. Soong, D. Atkinson, and D. Gehlert, "Analysis of a segmented brushless pm machine utilising soft magnetic composites", in *Industrial Electronics Society, 2007. IECON 2007. 33rd Annual Conference of the IEEE*, Nov. 2007, pp. 1268–1273.

[65] A. Jack, B. Mecrow, and C. Maddison, "Combined radial and axial permanent magnet motors using soft magnetic composites", in *Electrical Machines and Drives, 1999. Ninth International Conference on (Conf. Publ. No. 468)*, 1999, pp. 25–29.

[66] A. Reinap and M. Alaküla, "Impact of soft magnetic material on construction of radial flux electrical machines", *Magnetics, IEEE Transactions on*, vol. 48, no. 4, pp. 1613–1616, Apr. 2012.

[67] A. Jack, B. Mecrow, P. Dickinson, D. Stephenson, J. Burdess, N. Fawcett, and J. Evans, "Permanent-magnet machines with powdered iron cores and prepressed windings", *Industry Applications, IEEE Transactions on*, vol. 36, no. 4, pp. 1077–1084, Jul. 2000.

[68] H. Amano, Y. Enomoto, M. Ito, R. Masaki, M. Masuzawa, and M. Mita, "Characteristics of a permanent-magnet synchronous motor with a dual-molding permanent-magnet rotor", in *Power Engineering Society General Meeting, 2007. IEEE*, Jun. 2007, pp. 1–7.

[69] N. Bianchi, S. Bolognani, and P. Frare, "Design criteria of high efficiency spm synchronous motors", in *Electric Machines and Drives Conference, 2003. IEMDC'03. IEEE International*, vol. 2, Jun. 2003, pp. 1042–1048.

[70] J. Wang and D. Howe, "Influence of soft magnetic materials on the design and performance of tubular permanent magnet machines", *IEEE TRANSACTIONS ON MAGNETICS*, vol. 41, 2005.

[71] T. Kosaka, Y. Kano, N. Matsui, and C. Pollock, "A novel multi-pole permanent magnet synchronous machine with smc bypass core for magnet flux and smc field-pole core with toroidal coil for independent field strengthening/weakening", in *Power Electronics and Applications, 2005 European Conference on*, Sep. 2005, pp. 1–10.

[72] G. Cvetkovski and L. Petkovska, "Performance improvement of pm synchronous motor by using soft magnetic composite material",

Magnetics, IEEE Transactions on, vol. 44, no. 11, pp. 3812–3815, Nov. 2008.

[73] T. Henneron, S. Clénet, J. Cros, and P. Viarouge, "Evaluation of 3-d finite element method to study and design a soft magnetic composite machine", *Magnetics, IEEE Transactions on*, vol. 40, no. 2, pp. 786–789, Mar. 2004.

[74] J. Cros, P. Viarouge, and M. Kakhki, "Design and optimization of soft magnetic composite machines with finite element methods", *Magnetics, IEEE Transactions on*, vol. 47, no. 10, pp. 4384–4390, Oct. 2011.

[75] T. Ishikawa, K. Takahashi, Q. V. Ho, M. Matsunami, and N. Kurita, "Analysis of novel brushless dc motors made of soft magnetic composite core", *Magnetics, IEEE Transactions on*, vol. 48, pp. 971–974, Feb. 2012.

[76] T. Ishikawa, Y. Sato, and N. Kurita, "Performance of novel permanent magnet synchronous machines made of soft magnetic composite core", *Magnetics, IEEE Transactions on*, vol. 50, no. 11, pp. 1–4, Nov. 2014.

[77] M. Aydin, S. Huang, and T. Lipo, "Axial flux permanent magnet disc machines: A review", *Proceedings of the 2004 SPEEDAM*, 2004.

[78] J. Gieras, R. Wang, and M. Kamper, *Axial flux permanent magnet brushless machines*. Springer Science & Business Media, 2008.

[79] G. S. Liew, "Analysis and design of single-sided, slotted amm axial-field permanent magnet machines.", PhD thesis, School of Electric Engineering, University of Adelaide, Australia, 2009.

[80] A. . Mahmoudi, N. Rahim, and W. Hew, "Axial flux permanent magnet machine modeling, design, simulation, and analysis", *Scientific Research and Essays*, vol. 6, no. 12, pp. 2525–2549, 2011.

[81] S. Kahourzade, A. Mahmoudi, W. Hew, and M. Uddin, "A comprehensive review of axial-flux permanent magnet machines", *Electrical and Computer Engineering, Canadian Journal of*, vol. 37, no. 1, pp. 19–33, Oct. 2014.

[82] M. Sanada, S. Morimoto, and Y. Takeda, "Axial type flat pm motor with large air gap", in *Power Conversion Conference - Nagaoka 1997., Proceedings of the*, vol. 2, Aug. 1997, pp. 643–647.

[83] C. Liu and K. Chuang, "On the design of a disc-type surface-mounted permanent magnet motor for electric scooter application", in *Industry Applications Conference, 2002. 37th IAS Annual Meeting. Conference Record of the*, vol. 1, Oct. 2002, pp. 377–383.

[84] R. NILSSEN, S. SKAAR, R. LUND, T. SKJELLNES, S. OVREBO, and E. LOVLI, “Design of a permanent magnet synchronous motor integrated in the wheel rim on wheelchairs”, in *Power Electronics and Applications, 2005 European Conference on*, Sep. 2005, pp. 1–8.

[85] D. PATTERSON and R. SPEE, “The design and development of an axial flux permanent magnet brushless dc motor for wheel drive in a solar powered vehicle”, *Industry Applications, IEEE Transactions on*, vol. 31, no. 5, pp. 1054–1061, Sep. 1995.

[86] G. LIEW, N. ERTUGRUL, W. SOONG, and D. GEHLERT, “Analysis and performance evaluation of an axial-field brushless pm machine utilising soft magnetic composites”, in *Electric Machines Drives Conference, 2007. IEMDC ’07. IEEE International*, vol. 1, May 2007, pp. 153–158.

[87] F. MARIGNETTI, V. COLLI, R. DI STEFANO, and A. CAVAGNINO, “Design issues of a fractional-slot windings axial flux pm machine with soft magnetic compound stator”, in *Industrial Electronics Society, 2007. IECON 2007. 33rd Annual Conference of the IEEE*, Nov. 2007, pp. 187–192.

[88] Z. ZHANG, F. PROFUMO, A. TENCONI, and M. SANTAMARIA, “Analysis and experimental validation of performance for an axial flux permanent magnet brushless dc motor with powder iron metallurgy cores”, *Magnetics, IEEE Transactions on*, vol. 33, no. 5, pp. 4194–4196, Sep. 1997.

[89] S.-U. CHUNG, G.-Y. HWANG, S.-M. HWANG, B.-S. KANG, and H. gyoo KIM, “Development of brushless and sensorless vibration motor used for mobile phone”, *Magnetics, IEEE Transactions on*, vol. 38, no. 5, pp. 3000–3002, Sep. 2002.

[90] T. CHAN and L. LAI, “An axial-flux permanent-magnet synchronous generator for a direct-coupled wind-turbine system”, *Energy Conversion, IEEE Transactions on*, vol. 22, no. 1, pp. 86–94, Mar. 2007.

[91] T. CHAN, L. LAI, and S. XIE, “Field computation for an axial flux permanent-magnet synchronous generator”, *Energy Conversion, IEEE Transactions on*, vol. 24, no. 1, pp. 1–11, Mar. 2009, ISSN: 0885-8969.

[92] Y.-P. YANG, Y.-P. LUH, and C.-H. CHEUNG, “Design and control of axial-flux brushless dc wheel motors for electric vehicles-part i: Multiobjective optimal design and analysis”, *Magnetics, IEEE Transactions on*, vol. 40, no. 4, pp. 1873–1882, Jul. 2004.

[93] Y.-P. YANG, J.-P. WANG, S.-W. WU, and Y.-P. LUH, “Design and control of axial-flux brushless dc wheel motors for electric vehicles-part

ii: Optimal current waveforms and performance test", *Magnetics, IEEE Transactions on*, vol. 40, no. 4, pp. 1883–1891, Jul. 2004.

[94] Y.-P. YANG and D. S. CHUANG, "Optimal design and control of a wheel motor for electric passenger cars", *Magnetics, IEEE Transactions on*, vol. 43, no. 1, pp. 51–61, Jan. 2007.

[95] F. PROFUMO, A. TENCONI, Z. ZHANG, and A. CAVAGNINO, "Novel axial flux interior pm synchronous motor realized with powdered soft magnetic materials", in *Industry Applications Conference, 1998. Thirty-Third IAS Annual Meeting. The 1998 IEEE*, vol. 1, Oct. 1998, pp. 152–158.

[96] M. KHAN, P. PILLAY, N. BATANE, and D. MORRISON, "Prototyping a composite smc/steel axial-flux pm wind generator", in *Industry Applications Conference, 2006. 41st IAS Annual Meeting. Conference Record of the 2006 IEEE*, vol. 5, Oct. 2006, pp. 2374–2381.

[97] F. CARICCHI, F. CRESCIMBINI, E. FEDELI, and G. NOIOA, "Design and construction of a wheel-directly-coupled axial-flux pm motor prototype for evs", in *Industry Applications Society Annual Meeting, 1994., Conference Record of the 1994 IEEE*, vol. 1, Oct. 1994, pp. 254–261.

[98] F. CARICCHI, F. CRESCIMBINI, O. HONORATI, A. DI NAPOLI, and E. SANTINI, "Compact wheel direct drive for evs", *Industry Applications Magazine, IEEE*, vol. 2, no. 6, pp. 25–32, Nov. 1996.

[99] P. ACARNLEY, B. MECROW, J. BURDESS, J. FAWCETT, J. KELLY, and P. DICKINSON, "Design principles for a flywheel energy store for road vehicles", *Industry Applications, IEEE Transactions on*, vol. 32, no. 6, pp. 1402–1408, Nov. 1996.

[100] Y. CHEN and P. PILLAY, "Axial-flux pm wind generator with a soft magnetic composite core", in *Industry Applications Conference, 2005. Fourtieth IAS Annual Meeting. Conference Record of the 2005*, vol. 1, Oct. 2005, pp. 231–237.

[101] M. KHAN, Y. CHEN, and P. PILLAY, "Application of soft magnetic composites to pm wind generator design", in *Power Engineering Society General Meeting, 2006. IEEE*, vol. 3, 2006, pp. 1–4.

[102] T. WOOLMER and M. MCCULLOCH, "Analysis of the yokeless and segmented armature machine", in *Electric Machines Drives Conference, 2007. IEMDC '07. IEEE International*, vol. 1, May 2007, pp. 704–708.

[103] O. MALOBERTI, R. FIGUEREDO, C. MARCHAND, Y. CHOUA, D. CONDAMIN, L. KOBYLANSKI, and E. BOMME, "3-d-2-d dynamic magnetic modeling of an axial flux permanent magnet motor with soft mag-

netic composites for hybrid electric vehicles", *Magnetics, IEEE Transactions on*, vol. 50, no. 6, pp. 1–11, Jun. 2014, ISSN: 0018-9464.

[104] S. HO, S. NIU, and W. FU, "Design and analysis of a novel axial-flux electric machine", *Magnetics, IEEE Transactions on*, vol. 47, no. 10, pp. 4368–4371, Oct. 2011.

[105] C. VERSÈLE, Z. DE GREVE, F. VALLEE, R. HANUISE, O. DEBLECKER, M. DELHAYE, and J. LOBRY, "Analytical design of an axial flux permanent magnet in-wheel synchronous motor for electric vehicle", in *Power Electronics and Applications, 2009. EPE '09. 13th European Conference on*, Sep. 2009, pp. 1–9.

[106] K. RAHMAN, N. PATEL, T. WARD, J. NAGASHIMA, F. CARICCHI, and F. CRESCIMBINI, "Application of direct-drive wheel motor for fuel cell electric and hybrid electric vehicle propulsion system", *Industry Applications, IEEE Transactions on*, vol. 42, no. 5, pp. 1185–1192, Sep. 2006.

[107] W. FEI, "Permanent magnet synchronous machines with fractional slot and concentrated winding configurations", PhD thesis, Defence Academy - College of Management and Technology, CRANFIELD UNIVERSITY, England, 2011.

[108] Z. WANG, R. MASAKI, S. MORINAGA, Y. ENOMOTO, H. ITABASHI, M. ITO, and S. TANIGAWA, "Development of an axial gap motor with amorphous metal cores", *Industry Applications, IEEE Transactions on*, vol. 47, no. 3, pp. 1293–1299, May 2011.

[109] Z. WANG, Y. ENOMOTO, H. TOKOI, A. KOMURA, T. OBATA, and K. SOMA, "Development of ie5 high efficiency motor with iron-base amorphous magnetic cores", in *Eenergy efficiency in motor driven systems Eemods15, 2015 International Conference on*, May 2015.

[110] G. DONATO, F. CAPPONI, and F. CARICCHI, "Fractional-slot concentrated-winding axial-flux permanent-magnet machine with core-wound coils", *Industry Applications, IEEE Transactions on*, vol. 48, no. 2, pp. 630–641, Mar. 2012.

[111] F. CARICCHI, F. CRESCIMBINI, and E. SANTINI, "Basic principle and design criteria of axial-flux pm machines having counterrotating rotors", *Industry Applications, IEEE Transactions on*, vol. 31, no. 5, pp. 1062–1068, Sep. 1995.

[112] F. CARICCHI, F. CRESCIMBINI, and O. HONORATI, "Low-cost compact permanent magnet machine for adjustable-speed pump application", *Industry Applications, IEEE Transactions on*, vol. 34, no. 1, pp. 109–116, Jan. 1998.

[113] F. Caricchi, F. Crescimbini, E. Santini, and L. Solero, "High-efficiency low-volume starter/alternator for automotive applications", in *Industry Applications Conference, 2000. Conference Record of the 2000 IEEE*, vol. 1, 2000, pp. 215–222.

[114] F. Crescimbini, A. Di Napoli, L. Solero, and F. Caricchi, "Compact permanent-magnet generator for hybrid vehicle applications", *Industry Applications, IEEE Transactions on*, vol. 41, no. 5, pp. 1168–1177, Sep. 2005.

[115] R. Ficheux, F. Caricchi, F. Crescimbini, and O. Honorati, "Axial-flux permanent-magnet motor for direct-drive elevator systems without machine room", *Industry Applications, IEEE Transactions on*, vol. 37, no. 6, pp. 1693–1701, Nov. 2001.

[116] F. Caricchi, F. Maradei, G. De Donato, and F. Capponi, "Axial-flux permanent-magnet generator for induction heating gensets", *Industrial Electronics, IEEE Transactions on*, vol. 57, pp. 128–137, Jan. 2010.

[117] B. Chalmers and E. Spooner, "An axial-flux permanent-magnet generator for a gearless wind energy system", *Energy Conversion, IEEE Transactions on*, vol. 14, pp. 251–257, Jun. 1999.

[118] F. Caricchi, F. Crescimbini, O. Honorati, G. Bianco, and E. Santini, "Performance of coreless-winding axial-flux permanent-magnet generator with power output at 400 hz, 3000 r/min", *Industry Applications, IEEE Transactions on*, vol. 34, no. 6, pp. 1263–1269, Nov. 1998.

[119] J. Bumby and R. Martin, "Axial-flux permanent-magnet air-cored generator for small-scale wind turbines", *Electric Power Applications, IEE Proceedings on*, vol. 152, pp. 1065–1075, Sep. 2005.

[120] J. Santiago, J. Oliveira, J. Lundin, A. Larsson, and H. Bernhoff, "Losses in axial-flux permanent-magnet coreless flywheel energy storage systems", in *Electrical Machines, 2008. ICEM 2008. 18th International Conference on*, Sep. 2008, pp. 1–5.

[121] H. Lovatt, V. Ramsden, and B. Mecrow, "Design of an in-wheel motor for a solar-powered electric vehicle", *Electric Power Applications, IEE Proceedings -*, vol. 145, no. 5, pp. 402–408, Sep. 1998.

[122] A. Di Gerlando, G. Foglia, M. Iacchetti, and R. Perini, "Axial flux pm machines with concentrated armature windings: Design analysis and test validation of wind energy generators", *Industrial Electronics, IEEE Transactions on*, vol. 58, no. 9, pp. 3795–3805, Sep. 2011.

[123] F. CARICCHI, F. CRESCIMBINI, and O. HONRATI, "Modular axial-flux permanent-magnet motor for ship propulsion drives", *Energy Conversion, IEEE Transactions on*, vol. 14, no. 3, pp. 673–679, Sep. 1999.

[124] F. LIBERT, "Design, optimization and comparison of permanent magnet motors for a low-speed direct-driven mixer", PhD thesis, Royal Institute of Technology, KTH, Stockholm, 2004.

[125] F. CARICCHI, F. CRESCIMBINI, and A. DI NAPOLI, "Prototype of innovative wheel direct drive with water-cooled axial-flux pm motor for electric vehicle applications", in *Applied Power Electronics Conference and Exposition, 1996. APEC '96. Conference Proceedings 1996., Eleventh Annual*, vol. 2, May 1996, pp. 764–770.

[126] F. CARICCHI, F. CRESCIMBINI, F. MEZZETTI, and E. SANTINI, "Multistage axial-flux pm machine for wheel direct drive", *Industry Applications, IEEE Transactions on*, vol. 32, pp. 882–888, Jul. 1996.

[127] F. CARICCHI, F. CRESCIMBINI, A. DI NAPOLI, and M. MARCHEGGIANI, "Prototype of electric vehicle drive with twin water-cooled wheel direct drive motors", in *Power Electronics Specialists Conference, 1996. PESC'96 Record., 27th Annual IEEE*, vol. 2, Jun. 1996, pp. 1926–1932.

[128] J. EASTHAM, F. PROFUMO, A. TENCONI, R. HILL-COTTINGHAM, P. COLES, and G. GIANOLIO, "Novel axial flux machine for aircraft drive: Design and modeling", *Magnetics, IEEE Transactions on*, vol. 38, no. 5, pp. 3003–3005, Sep. 2002.

[129] S. HUANG, J. LUO, F. LEONARDI, and T. LIPO, "A comparison of power density for axial flux machines based on general purpose sizing equations", *Energy Conversion, IEEE Transactions on*, vol. 14, no. 2, pp. 185–192, Jun. 1999, ISSN: 0885-8969.

[130] K. SITAPATI and R. KRISHNAN, "Performance comparisons of radial and axial field, permanent-magnet, brushless machines", *Industry Applications, IEEE Transactions on*, vol. 37, no. 5, pp. 1219–1226, Sep. 2001.

[131] A. CAVAGNINO, M. LAZZARI, F. PROFUMO, and A. TENCONI, "A comparison between the axial flux and the radial flux structures for pm synchronous motors", *Industry Applications, IEEE Transactions on*, vol. 38, no. 6, pp. 1517–1524, Nov. 2002.

[132] K AKATSU and S WAKUI, "A comparison between axial and radial flux pm motor by optimum design method from the required output nt characteristics", *COMPEL-The international journal for computation*

and mathematics in electrical and electronic engineering, vol. 25, no. 2, pp. 496–509, 2006.

[133] Z. ZHU and D. HOWE, “Instantaneous magnetic field distribution in brushless permanent magnet dc motors. iii. effect of stator slotting”, *Magnetics, IEEE Transactions on*, vol. 29, no. 1, pp. 143–151, Jan. 1993, ISSN: 0018-9464.

[134] Y. ZHILICHEV, “Three-dimensional analytic model of permanent magnet axial flux machine”, *Magnetics, IEEE Transactions on*, vol. 34, pp. 3897–3901, 1998.

[135] H. TIEGNA, Y. AMARA, and G. BARAKAT, “A new quasi-3-d analytical model of axial flux permanent magnet machines”, *Magnetics, IEEE Transactions on*, vol. 50, 2014.

[136] Y. KANO, T. KOSAKA, and N. MATSUI, “A simple nonlinear magnetic analysis for axial-flux permanent-magnet machines”, *Industrial Electronics, IEEE Transactions on*, vol. 57, pp. 2124–2133, Jun. 2010.

[137] C. BOCCALETTI, C. BRUZZESE, E. SANTINI, and P. SORDI, “Accurate design of axial-flux permanent-magnet synchronous machines by means of a 3d finite element analysis”, *Proceedings of the 5-th ELECTROMOTION, International Symposium on 'Advanced Electromechanical Motion Systems',*, vol. 10, no. 3, pp. 223–228, 2003.

[138] A. PARVIAINEN, M. NIEMELA, and J. PYRHONEN, “Modeling of axial flux permanent-magnet machines”, *Industry Applications, IEEE Transactions on*, vol. 40, pp. 1333–1340, 2004.

[139] J. AZZOUZI, G. BARAKAT, and B. DAKYO, “Quasi-3-d analytical modeling of the magnetic field of an axial flux permanent-magnet synchronous machine”, *Energy Conversion, IEEE Transactions on*, vol. 20, pp. 746–752, 2005.

[140] S. HUANG, M. AYDIN, and T. LIPO, “Comparison of (non-slotted and slotted) surface mounted pm motors and axial flux motors for submarine ship drives”, in *Third Naval Symposium on Electrical Machines, Philadelphia*, 2000.

[141] M. AYDIN, S. HUANG, and T. LIPO, “Torque quality and comparison of internal and external rotor axial flux surface-magnet disc machines”, *Industrial Electronics, IEEE Transactions on*, vol. 53, no. 3, pp. 822–830, Jun. 2006.

[142] W. FEI and P. LUK, “Performance study of two axial-flux permanent-magnet machine topologies with soft magnetic composite cores”, in *Power Electronics and Motion Control Conference, 2009. IPEMC '09. IEEE 6th International*, May 2009, pp. 411–417.

[143] A. MAHMOUDI, H. PING, and N. RAHIM, “A comparison between the torus and afir axial-flux permanent-magnet machine using finite element analysis”, in *Electric Machines Drives Conference (IEMDC), 2011 IEEE International*, May 2011, pp. 242–247.

[144] T. WOOLMER and M. MCCULLOCH, “Axial flux permanent magnet machines: A new topology for high performance applications”, *Hybrid Vehicle Conference (Conf. Pub. CP526), IET*, pp. 27–42, 2006.

[145] S HUANG, M. AYDIN, and T. LIPO, “Torus concept machines: Pre-prototyping design assessment for two major topologies”, in *Industry Applications Conference, 2001. Thirty-Sixth IAS Annual Meeting. Conference Record of the 2001 IEEE*, vol. 3, Sep. 2001, pp. 1619–1625.

[146] M. AYDIN, S. HUANG, and T. LIPO, “Design and 3d electromagnetic field analysis of non-slotted and slotted torus type axial flux surface mounted permanent magnet disc machines”, in *Electric Machines and Drives Conference, 2001. IEMDC 2001. IEEE International*, 2001, pp. 645–651.

[147] F. SAHIN, “Design and development of a high-speed axial-flux permanent-magnet machine”, PhD thesis, Technische Universiteit Eindhoven, 2001.

[148] S. LOMHEIM, “Analysis of a novel coil design for axial flux machines”, Master’s thesis, Norwegian University of Science and Technology, 2013.

[149] M. AYDIN, S. HUANG, and T. LIPO, “A new axial flux surface mounted permanent magnet machine capable of field control”, in *Industry Applications Conference, 2002. 37th IAS Annual Meeting. Conference Record of the*, vol. 2, Oct. 2002, pp. 1250–1257.

[150] M AYDIN, J YAO, E KAYIKCI, S HUANG, and T. LIPO, “Design considerations and experimental results of an axial flux pm motor with field control”, in *Proceedings of ICEM*, vol. 4, 2004, pp. 764–770.

[151] F. DUMAS, D. MATT, J. JAC, and P. ENRICI, “A design approach for an axial-flux permanent magnet motor”, in *Electrical Machines (ICEM), 2010 XIX International Conference on*, Sep. 2010, pp. 1–6.

[152] J. BRENNVALL, “Coil assembly for three phased transverse axial flux multi disk machines”, pat. WO 2012128646 A1, 2012.

[153] M.-F. HSIEH, D. DORRELL, Y.-H. YEH, and S. EKRAM, “Cogging torque reduction in axial flux machines for small wind turbines”, in *Industrial Electronics, 2009. IECON ’09. 35th Annual Conference of IEEE*, Nov. 2009, pp. 4435–4439.

[154] D. GONZALEZ, J. TAPIA, and A. BETTANCOURT, “Design consideration to reduce cogging torque in axial flux permanent-magnet machines”, *Magnetics, IEEE Transactions on*, vol. 43, no. 8, pp. 3435–3440, Aug. 2007.

[155] M. AYDIN, R. QU, and T. LIPO, “Cogging torque minimization technique for multiple-rotor, axial-flux, surface-mounted-pm motors: Alternating magnet pole-arcs in facing rotors”, in *Industry Applications Conference, 2003. 38th IAS Annual Meeting. Conference Record of the*, vol. 1, Oct. 2003, pp. 555–561.

[156] A. B. LETELIER, D. A. GONZALEZ, J. A. TAPIA, R. WALLACE, and M. A. VALENZUELA, “Cogging torque reduction in an axial flux pm machine via stator slot displacement and skewing”, *IEEE Transactions on Industry Applications*, vol. 43, no. 3, pp. 685–693, May 2007.

[157] Z. Q. ZHU and D. HOWE, “Influence of design parameters on cogging torque in permanent magnet machines”, *IEEE Transactions on Energy Conversion*, vol. 15, no. 4, pp. 407–412, Dec. 2000.

[158] F. CARICCHI, F. CAPPONI, F. CRESCIMBINI, and L. SOLERO, “Experimental study on reducing cogging torque and no-load power loss in axial-flux permanent-magnet machines with slotted winding”, *Industry Applications, IEEE Transactions on*, vol. 40, no. 4, pp. 1066–1075, Jul. 2004.

[159] H WEH and H MAY, “Achievable force densities for permanent magnet excited machines in new configurations”, in *Proc. Int. Conf. on Electrical Machines*, 1986, pp. 1107–1111.

[160] H WEH, H HOFFMAN, and J LANDRATH, “New permanent magnet excited synchronous machine with high efficiency at low speeds”, in *Proc. Int. Conf. Electrical Machines*, vol. 3, 1988, pp. 35–40.

[161] H WEH, “Permanentmagneterregte synchronmaschinen hoher kraftdichte nach dem transversalflusskonzept”, *Etz-Archiv*, vol. 10, no. 5, pp. 143–149, 1988.

[162] I. VIOREL, M. CRIVII, L. LÖWENSTEIN, L. SZABÓ, and M. GUTMAN, “Direct drive systems with transverse flux reluctance motors”, *Acta Electrotehnica*, vol. 45, no. 3, pp. 33–40, 2004.

[163] G. HENNENBERGER, L. LÖWENSTEIN, and L. .ZACHARIAS, “Crankshaft starter generators based on the principle of reluctance”, *AutoTechnology*, vol. 2, no. 6, pp. 68–71, 2002.

[164] R. KRUSE, G. PFAFF, and C. PFEIFFER, “Transverse flux reluctance motor for direct servodrive applications”, in *Industry Applications Con-*

ference, 1998. Thirty-Third IAS Annual Meeting. The 1998 IEEE, vol. 1, Oct. 1998, pp. 655–662.

[165] D. POPA, V. IANCU, and L. SZABÓ, “Linear transverse flux motor for conveyors”, in *6th International Symposium on Linear Drives for Industrial Application (LDIA’2007), Lille, France, paper*, vol. 188, 2007.

[166] D. POPA, V. IANCU, and L. SZABO, “Improved design of a linear transverse flux reluctance motor”, in *Optimization of Electrical and Electronic Equipment, 2008. OPTIM 2008. 11th International Conference on*, May 2008, pp. 137–142.

[167] H. AMREIZ, “A comparison between transverse flux and conventional switched reluctance machines”, in *Electrical Machines (ICEM), 2010 XIX International Conference on*, Sep. 2010, pp. 1–8.

[168] J. DOERING, G. STEINBORN, and W. HOFMANN, “Torque, power, losses and heat calculation of a transverse flux reluctance machine with soft magnetic composite materials and disc-shaped rotor”, in *Energy Conversion Congress and Exposition (ECCE), 2013 IEEE*, Sep. 2013, pp. 4326–4333.

[169] ——, “Torque, power, losses, and heat calculation of a transverse flux reluctance machine with soft magnetic composite materials and disk-shaped rotor”, *Industry Applications, IEEE Transactions on*, vol. 51, no. 2, pp. 1494–1504, Mar. 2015.

[170] P. RASMUSSEN, G. RUNOLFSSON, T. THORSDOTTIR, U. JAKOBSEN, and A. PEDERSEN, “E-core transverse flux machine with integrated fault detection system”, in *Electrical Machines and Systems (ICEMS), 2011 International Conference on*, Aug. 2011, pp. 1–6.

[171] P. BOLOGNESI, “Design and manufacturing of an unconventional variable reluctance machine”, in *Power Electronics, Machines and Drives, 2008. PEMD 2008. 4th IET Conference on*, Apr. 2008, pp. 45–49.

[172] A. COSIC, C. SADARANGANI, and J. TIMMERMAN, “Design and manufacturing of a linear transverse flux permanent magnet machines”, in *Industry Applications Society Annual Meeting, 2008. IAS ’08. IEEE*, Oct. 2008, pp. 1–5.

[173] A. COSIC, “Analysis of a novel transversal flux machine with a tubular cross-section for free piston energy converter application”, PhD thesis, Kungl Tekniska Högskolan (KTH), Sweden, 2010.

[174] G. HENNEBERGER and M. BORK, “Development of a new transverse flux motor”, in *New Topologies for Permanent Magnet Machines, IEE Colloquium on*, vol. 1, Jun. 1997, pp. 1–6.

[175] D. SVECHKARENKO, J. SOULARD, and C. SADARANGANI, "A novel transverse flux generator in direct-driven wind turbines", in *Proc. Nordic Workshop on Power and Industrial Electronics*, 2006.

[176] W. ARSHAD, T. BACKSTROM, and C. SADARANGANI, "Analytical design and analysis procedure for a transverse flux machine", in *Electric Machines and Drives Conference, 2001. IEMDC 2001. IEEE International*, 2001, pp. 115–121.

[177] J. GIERAS, "Performance characteristics of a transverse flux generator", in *Electric Machines and Drives, 2005 IEEE International Conference on*, May 2005, pp. 1293–1299.

[178] J. LEE, D. HONG, B. WOO, D. PARK, and B. NAM, "Performance comparison of longitudinal flux and transverse flux permanent magnet machines for turret applications with large diameter", *Magnetics, IEEE Transactions on*, vol. 48, no. 2, pp. 915–918, Feb. 2012.

[179] A. CHEN, R. NILSSEN, and A. NYSVEEN, "Performance comparisons among radial-flux, multistage axial-flux, and three-phase transverse-flux pm machines for downhole applications", *IEEE Transactions on Industry Applications*, vol. 46, no. 2, pp. 779–789, Mar. 2010.

[180] Y. GUO, J. ZHU, P. WATTERSON, and W. WU, "Development of a pm transverse flux motor with soft magnetic composite core", *Energy Conversion, IEEE Transactions on*, vol. 21, no. 2, pp. 426–434, 2006.

[181] Y. GUO, J. ZHU, J. CHEN, and J. JIN, "Performance analysis of a pm claw pole smc motor with brushless dc control scheme", in *Power Electronics and Motion Control Conference, 2006. IPEMC 2006. CES/IEEE 5th International*, vol. 2, Aug. 2006, pp. 1–5.

[182] Y. HUANG, J. ZHU, Y. GUO, Z. LIN, and Q. HU, "Design and analysis of a high-speed claw pole motor with soft magnetic composite core", *Magnetics, IEEE Transactions on*, vol. 43, no. 6, pp. 2492–2494, Jun. 2007.

[183] Y. HUANG, Q. HU, J. ZHAO, J. ZHU, and Y. GUO, "Comparative study of high-speed pm motors with laminated steel and soft magnetic composite cores", in *Industry Applications Conference, 2007. 42nd IAS Annual Meeting. Conference Record of the 2007 IEEE*, Sep. 2007, pp. 67–72.

[184] Y. HUANG, J. ZHU, Y. GUO, and Q. HU, "Development of a high-speed claw pole motor with soft magnetic composite core", in *Electric Machines Drives Conference, 2007. IEMDC 2007. IEEE International Conference on*, vol. 2, May 2007, pp. 1564–1568.

[185] F. ZHANG, Y. LIU, and H. BAI, "Study and design of 3-d flux permanent magnet claw-pole motor with smc stator", in *Applied Superconductivity and Electromagnetic Devices, 2009. ASEMD 2009. International Conference on*, Sep. 2009, pp. 316–319.

[186] S. HOSSEINI, J. MOGHANI, N. ERSHAD, and B. JENSEN, "Design, prototyping, and analysis of a novel modular permanent-magnet transverse flux disk generator", *Magnetics, IEEE Transactions on*, vol. 47, no. 4, pp. 772–780, Apr. 2011.

[187] P. ANPALAHAN, "Design of transverse flux machines using analytical calculations & finite element analysis", PhD thesis, Kungl Tekniska Högskolan, Sweden, 2001.

[188] J. CHENG, X. TU, and G. C., "Experimental research on new-type of transverse flux permanent magnet machine for electric vehicle", in *Electrical Machines and Systems, 2008. ICEMS 2008. International Conference on*, Oct. 2008, pp. 3220–3223.

[189] G. PATTERSON, T. KOSEKI, Y. AOYAMA, and K. SAKO, "Simple modeling and prototype experiments for a new high-thrust low-speed permanent-magnet disk motor", *Industry Applications, IEEE Transactions on*, vol. 47, no. 1, pp. 65–71, Jan. 2011.

[190] M. S. WIDYAN, "Design, optimization, construction and test of rare-earth permanent-magnet electrical machines with new topology for wind energy applications", PhD thesis, Technische Universität Berlin, 2006.

[191] D. BANG, H. POLINDER, G. SHRESTHA, and J. FERREIRA, "Design of a lightweight transverse flux permanent magnet machine for direct-drive wind turbines", in *Industry Applications Society Annual Meeting, 2008. IAS '08. IEEE*, Oct. 2008, pp. 1–7.

[192] A. NJEH, A. MASMOUDI, and A. ELANTABLY, "3d fea based investigation of the cogging torque of a claw pole transverse flux permanent magnet machine", in *Electric Machines and Drives Conference, 2003. IEMDC'03. IEEE International*, vol. 1, Jun. 2003, pp. 319–324.

[193] D. KANG, Y. CHUN, and H. WEH, "Analysis and optimal design of transverse flux linear motor with pm excitation for railway traction", *Electric Power Applications, IEE Proceedings on*, vol. 150, no. 4, pp. 493–499, Jul. 2003.

[194] E. SCHMIDT, D. BRUNNSCHWEILER, and S. BERCHTEN, "Electromagnetic parameters of a three phase transverse flux machine with an external rotor for wheel hub drives", in *Universities Power Engineering Conference (AUPEC), 2010 20th Australasian*, Dec. 2010, pp. 1–6.

[195] H. POLINDER, B. C. MECROW, A. G. JACK, P. G. DICKINSON, and M. A. MUELLER, "Conventional and tfpm linear generators for direct-drive wave energy conversion", *IEEE Transactions on Energy Conversion*, vol. 20, pp. 260–267, Jun. 2005.

[196] A. MITCHAM, "Transverse flux motors for electric propulsion of ships", in *New Topologies for Permanent Magnet Machines, IEE Colloquium on*, vol. 3, Jun. 1997, pp. 1–6.

[197] S. HUSBAND and C. HODGE, "The rolls-royce transverse flux motor development", in *Electric Machines and Drives Conference, 2003. IEMDC'03. IEEE International*, vol. 3, Jun. 2003, pp. 1435–1440.

[198] A. MASMOUDI and A. ELANTABLY, "A simple assessment of the cogging torque in a transverse flux permanent magnet machine", in *Electric Machines and Drives Conference, 2001. IEMDC 2001. IEEE International*, 2001, pp. 754–759.

[199] J. CHANG, D. KANG, J LEE, and J. HONG, "Development of transverse flux linear motor with permanent-magnet excitation for direct drive applications", *Magnetics, IEEE Transactions on*, vol. 41, no. 5, pp. 1936–1939, May 2005.

[200] E. SCHMIDT, "3-d finite element analysis of the cogging torque of a transverse flux machine", *Magnetics, IEEE Transactions on*, vol. 41, no. 5, pp. 1836–1839, May 2005.

[201] A. MASMOUDI and A. ELANTABLY, "An approach to sizing high power density tfpm intended for hybrid bus electric propulsion", *Electric Machines &Power Systems*, vol. 28, no. 4, pp. 341–354, 2000.

[202] H. GROKE, J. SCHÜTTLER, J. ADLER, and B. ORLIK, "Optimized power-feedback with an 8.7 knm-transverse flux generator", in *Electrical Machines (ICEM), 2010 XIX International Conference on*, Sep. 2010, pp. 1–6.

[203] Y. GONG, W. ZHENG, D. ZHANG, and J. JIANG, "Analysis of a transverse flux machine with e-shaped stator using three dimensional scalar potential finite element method", in *Power Electronics and Motion Control Conference, 2009. IPEMC '09. IEEE 6th International*, May 2009, pp. 1991–1994.

[204] M. R. HARRIS, G. H. PAJOOMAN, and S. M. ABU SHARKH, "Performance and design optimisation of electric motors with heteropolar surface magnets and homopolar windings", *IEE Proceedings - Electric Power Applications*, vol. 143, no. 6, pp. 429–436, Nov. 1996, ISSN: 1350-2352.

[205] J. PIPPURI, A. MANNINEN, J. KERÄNEN, and K. TAMMI, "Torque density of radial, axial and transverse flux permanent magnet machine topologies", *IEEE Transactions on Magnetics*, vol. 49, no. 5, pp. 2339–2342, May 2013.

[206] M. GARTNER, P. SEIBOLD, and N. PARSPOUR, "Laminated circumferential transverse flux machines-lamination concept and applicability to electrical vehicles", in *Electric Machines Drives Conference (IEMDC), 2011 IEEE International*, May 2011, pp. 831–837.

[207] J. CHANG, D. KANG, I. VIOREL, and S. LARISA, "Transverse flux reluctance linear motor's analytical model based on finite-element method analysis results", *Magnetics, IEEE Transactions on*, vol. 43, no. 4, pp. 1201–1204, Apr. 2007.

[208] Y. JEONG, D. KANG, J. KIM, and S. JANG, "A design of transverse flux motor with permanent magnet shield", in *Industrial Electronics, 2001. Proceedings. ISIE 2001. IEEE International Symposium on*, vol. 2, 2001, pp. 995–999.

[209] D. KANG, Y. JEONG, and M. KIM, "A study on the design of transverse flux linear motor with high power density", in *Industrial Electronics, 2001. Proceedings. ISIE 2001. IEEE International Symposium on*, vol. 2, 2001, pp. 707–711.

[210] D. KANG, "Increasing of thrust force in transverse flux machine by permanent-magnet screen", *Magnetics, IEEE Transactions on*, vol. 41, no. 5, pp. 1952–1955, May 2005.

[211] H. YANG and C. GU, "Analytical design and modeling of transverse flux switched reluctance machine", in *Electrical Machines and Systems, 2008. ICEMS 2008. International Conference on*, Oct. 2008, pp. 3414–3416.

[212] F. DREHER and N. PARSPOUR, "A novel high-speed permanent magnet claw pole transverse flux machine for use in automation", in *Power Electronics, Electrical Drives, Automation and Motion (SPEEDAM), 2012 International Symposium on*, Jun. 2012, pp. 1240–1245.

[213] Y. GUO, J. ZHU, and W. WU, "Design and analysis of electric motors with soft magnetic composite core", *Studies in Applied Electromagnetics and Mechanics*, vol. 15, no. 1, pp. 93–99, 2004.

[214] M. HARRIS, G. PAJOOMAN, and S. ABU SHARKH, "The problem of power factor in vrpm (transverse-flux) machines", in *Electrical Machines and Drives, 1997 Eighth International Conference on (Conf. Publ. No. 444)*, Sep. 1997, pp. 386–390.

[215] M. SIATKOWSKI and B. ORLIK, "Flux linkage in transverse flux machines with flux concentration", in *Optimization of Electrical and Electronic Equipment, 2008. OPTIM 2008. 11th International Conference on*, May 2008, pp. 21–26.

[216] S. BASERRAH and B. ORLIK, "Comparison study of permanent magnet transverse flux motors (pmtfms) for in-wheel applications", in *Power Electronics and Drive Systems, 2009. PEDS 2009. International Conference on*, Nov. 2009, pp. 96–101.

[217] M. HARRIS, G. PAJOOMAN, and S. SHARKH, "Comparison of alternative topologies for vrpm (transverse-flux) electrical machines", in *New Topologies for Permanent Magnet Machines (Digest No: 1997/090), IEE Colloquium on*, vol. 2, Jun. 1997, pp. 1–7.

[218] B. HASUBEK and E. NOWICKI, "Design limitations of reduced magnet material passive rotor transverse flux motors investigated using 3d finite element analysis", in *Electrical and Computer Engineering, 2000 Canadian Conference on*, vol. 1, 2000, pp. 365–369.

[219] B. E. HASUBEK, "Analysis and design of passive rotor transverse flux machines with permanent magnets on the stator", PhD thesis, University of Calgary, 2000.

[220] V. IANCU, T. CANTA, D. POPA, and L. SZABÓ, "Soft magnetic composites used for the iron core of the electrical machines", in *Proceedings of the 4th International Conference on Materials and Manufacturing Technologies, Cluj-Napoca, Romania*, 2006, pp. 125–129.

[221] K. LU, P. RASMUSSEN, and E. RITCHIE, "Design considerations of permanent magnet transverse flux machines", *Magnetics, IEEE Transactions on*, vol. 47, no. 10, pp. 2804–2807, Oct. 2011.

[222] P. ANPALAHAN, J. SOULARD, and H. NEE, "Design steps towards a high power factor transverse flux machine", in *Proc. European Conf. on Power Electronics and Applications*, 2001.

[223] D. SVECHKARENKO, J. SOULARD, and C. SADARANGANI, "Performance evaluation of a novel transverse flux generator with 3d finite element analysis", in *Electrical Machines and Systems, 2009. ICEMS 2009. International Conference on*, Nov. 2009, pp. 1–6.

[224] J. CHANG, J. LEE, J. KIM, S. CHUNG, D. KANG, and H. WEH, "Development of rotating type transverse flux machine", in *Electric Machines Drives Conference, 2007. IEMDC '07. IEEE International*, vol. 2, May 2007, pp. 1090–1095.

[225] S. BASERRAH, K. RIXEN, and B. ORLIK, "Transverse flux machines with distributed windings for in-wheel applications", in *Power Electron-*

ics and Drive Systems, 2009. PEDS 2009. International Conference on, Nov. 2009, pp. 102–108.

[226] S. BASERRAH, K. RIXEN, and B. ORLIK, "On construction of a compact in-wheel transverse flux machine with distributed windings", in *Power Conversion Intelligent Motion (PCIM2010), Proc. Int. Conf. and Exhibition for Power Electronics, Intelligent Motion, Power Quality*, 2010, pp. 580–586.

[227] M. DUBOIS, N. DEHLINGER, H. POLINDER, and D. MASSICOTTE, "Clawpole transverse-flux machine with hybrid stator", in *Electrical Machines,2006. ICEM 2006. 17th International Conference on*, 2006.

[228] J. LEE, S. MOON, D. KOO, D. KANG, G. LEE, and J. HONG, "Comparative study of stator core composition in transverse flux rotary machine", *Journal of Electrical Engineering & Technology*, vol. 6, no. 3, pp. 350–355, 2011.

[229] N. DEHLINGER and M. DUBOIS, "Clawpole transverse flux machines with amorphous stator cores", in *Electrical Machines, 2008. ICEM 2008. 18th International Conference on*, Sep. 2008, pp. 1–6.

[230] P. LEMIEUX, O. DELMA, M. DUBOIS, and R. GUTHRIE, "Soft magnetic composite with lamellar particles - application to the clawpole transverse-flux machine with hybrid stator", in *Electrical Machines, 2008. ICEM 2008. 18th International Conference on*, Sep. 2008, pp. 1–6.

[231] N. DEHLINGER and M. DUBOIS, "A new design method for the clawpole transverse flux machine. application to the machine no-load flux optimization. part 1: Accurate magnetic model with error compensation, part 2: Optimization aspects", in *Electrical Machines (ICEM), 2010 XIX International Conference on*, Sep. 2010, pp. 1–7.

[232] A. BABAZADEH, N. PARSPOUR, and A. HANIFI, "Transverse flux machine for direct drive robots: Modelling and analysis", in *Robotics, Automation and Mechatronics, 2004 IEEE Conference on*, vol. 1, Dec. 2004, pp. 376–380.

[233] Y. GUO, J. ZHU, and H. LU, "Accurate determination of parameters of a claw-pole motor with smc stator core by finite-element magnetic-field analysis", *Electric Power Applications, IEE Proceedings -*, vol. 153, no. 4, pp. 568–574, Jul. 2006.

[234] M. CRIVII, I. VIOREL, M. JUFER, and I. HUSAIN, "3d to 2d equivalence for a transverse flux reluctance motor", in *Proc. of ICEM*, vol. 2, 2002.

[235] J. Vining, T. Lipo, and G. Venkataramanan, "Design and optimization of a novel hybrid transverse / longitudinal flux, wound-field linear machine for ocean wave energy conversion", in *Energy Conversion Congress and Exposition, 2009. ECCE 2009. IEEE*, Sep. 2009, pp. 3726–3733.

[236] P. Zheng, Q. Zhao, J. Bai, B. Yu, Z. Song, and J. Shang, "Analysis and design of a transverse-flux dual rotor machine for power-split hybrid electric vehicle applications", *Energies*, vol. 6, no. 12, pp. 6548–6568, 2013.

[237] G. Kastinger, "Design of a novel transverse flux machine", in *Electrical Machines (ICEM), International Conference on*, 2002.

[238] Y. Hsu and M. Tsai, "Development of a novel transverse flux wheel motor", *Magnetics, IEEE Transactions on*, vol. 47, no. 10, pp. 3677–3680, Oct. 2011.

[239] M. Zafarani, M. Moallem, and A. Tabesh, "Analytical model for a transverse flux permanent magnet machine using improved magnetic equivalent circuit approach", in *Systems Engineering (ICSEng), 2011 21st International Conference on*, Aug. 2011, pp. 96–99.

[240] C. Espanet, J.-M. Kauffmann, and R. Bernard, "Comparison of two in-wheel permanent magnet motors for military applications", in *Vehicle Power and Propulsion Conference, 2006. VPPC '06. IEEE*, Sep. 2006, pp. 1–6.

[241] B. Zhang, "Messung und parameteridentifizierung von eisenverlusten fuer hoecherfrequente anwendungen", Master's thesis, RWTH Aachen, 2011.

[242] Y. Guo, J. Zhu, H. Lu, Z. Lin, S. Wang, and J. Jin, "3-d vector magnetic properties of smc material for advanced field analysis of smc machine", in *Power Engineering Conference, 2008. AUPEC '08. Australasian Universities*, Dec. 2008, pp. 1–4.

[243] Y. Guo, J. Zhu, H. Y. Lu, Z. Lin, and Y. Li, "Core loss calculation for soft magnetic composite electrical machines", *Magnetics, IEEE Transactions on*, vol. 48, no. 11, pp. 3112–3115, Nov. 2012.

[244] H. Barth, F. Bahr, N. Hildebrand, and W. Hofmann, "Einsatz weichmagnetischer pulververbundwerkstoffe in energieeffizienten transversalflussmaschinen", in *ETG-Fachbericht-Internationaler ETG-Kongress 2013–Energieversorgung auf dem Weg nach 2050*, VDE VERLAG GmbH, 2013.

[245] R. GOTTKEHASKAMP, "Optimal gefertigt systematischer entwurf von dreisträngigen zahnspulenwicklungen bürstenloser motoren", *Antriebstechnik*, pp. 1–6, 2007.

[246] R. GOTTKEHASKAMP and D. STRBAC, "Einsatz von zahnspulenwicklungen in asynchronmaschinen", *Antriebstechnik (ant Journal)*, vol. 3, pp. 26–30, 2013.

[247] F. LIBERT and J. SOULARD, "Investigation on pole-slot combinations for permanent-magnet machines with concentrated windings", in *International Conference on Electrical Machines, ICEM*, 2004, pp. 5–8.

[248] D. ISHAK, Z. ZHU, and D. HOWE, "Comparative study of permanent magnet brushless motors with all teeth and alternative teeth windings", in *Power Electronics, Machines and Drives, 2004. (PEMD 2004). Second International Conference on*, vol. 2, Mar. 2004, pp. 834–839.

[249] F. MAGNUSSEN and C. SADARANGANI, "Winding factors and joule losses of permanent magnet machines with concentrated windings", in *Electric Machines and Drives Conference, 2003. IEMDC'03. IEEE International*, vol. 1, Jun. 2003, pp. 333–339.

[250] J. CROS and P. VIAROUGE, "Synthesis of high performance pm motors with concentrated windings", *Energy Conversion, IEEE Transactions on*, vol. 17, no. 2, pp. 248–253, Jun. 2002.

[251] E. FORNASIERO, L. ALBERTI, N. BIANCHI, and S. BOLOGNANI, "Considerations on selecting fractional-slot nonoverlapped coil windings", *Industry Applications, IEEE Transactions on*, vol. 49, no. 3, pp. 1316–1324, May 2013.

[252] J. HENDERSHOT and T. MILLER, *Design of brushless permanent-magnet motors*. Magna Physics Pub., 1994.

[253] G. DAJAKU, W. XIE, and D. GERLING, "Reduction of low space harmonics for the fractional slot concentrated windings using a novel stator design", *Magnetics, IEEE Transactions on*, vol. 50, no. 5, pp. 1–12, May 2014.

[254] E. FORNASIERO, L. ALBERTI, N. BIANCHI, and S. BOLOGNANI, "Considerations on selecting fractional-slot windings", in *Energy Conversion Congress and Exposition (ECCE), 2010 IEEE*, Sep. 2010, pp. 1376–1383.

[255] N. BIANCHI, S. BOLOGNANI, M. PREÉ, and G. GREZZANI, "Design considerations for fractional-slot winding configurations of synchronous machines", *Industry Applications, IEEE Transactions on*, vol. 42, no. 4, pp. 997–1006, Jul. 2006.

[256] A. EL-REFAIE, "Fractional-slot concentrated-windings synchronous permanent magnet machines: Opportunities and challenges", *Industrial Electronics, IEEE Transactions on*, vol. 57, no. 1, pp. 107–121, Jan. 2010.

[257] M. HORZ, H. HERZOG, and A. HAAS, "Axial flux machine with single tooth fractional slot winding - comparison of different winding design approaches", in *Power Electronics, Electrical Drives, Automation and Motion, 2006. SPEEDAM 2006. International Symposium on*, May 2006, pp. 1444–1448.

[258] G. DAJAKU and D. GERLING, "A novel tooth concentrated winding with low space harmonic contents", in *Electric Machines Drives Conference (IEMDC), 2013 IEEE International*, May 2013, pp. 755–760.

[259] F. MAGNUSSEN, D. SVECHKARENKO, P. THELIN, and C. SADARANGANI, "Analysis of a pm machine with concentrated fractional pitch windings", *The proceedings of the Nordic Worshop on Power and Industrial Electronics (NORPIE), Trondheim (Norway)*, pp. 14–15, 2004.

[260] G. HUTH, "Permanent-magnet-excited ac servo motors in tooth-coil technology", *Energy Conversion, IEEE Transactions on*, vol. 20, no. 2, pp. 300–307, Jun. 2005.

[261] O. W., A. EL-ANTABLY, and T. LIPO, "Analysis of optimal stator concentric winding patterns design", in *Electrical Machines and Systems, 2005. ICEMS 2005. Proceedings of the Eighth International Conference on*, vol. 1, Sep. 2005, pp. 94–98.

[262] P. SRIKOMKHAM and S. RUANGSINCHAIWANICH, "Optimal rotor design of a psc motor using taguchi method and fem", in *Electrical Machines and Systems (ICEMS), 2010 International Conference on*, Oct. 2010, pp. 1341–1346.

[263] C. C. HWANG, L. Y. LYU, C. T. LIU, and P. L. LI, "Optimal design of an spm motor using genetic algorithms and taguchi method", *Magnetics, IEEE Transactions on*, vol. 44, pp. 4325–4328, Nov. 2008.

[264] U. KURT and G. ONBILGIN, "Design and optimization of axial flux permanent magnet synchronous machines using taguchi approach", in *Electrical and Electronics Engineering, 2009. ELECO 2009. International Conference on*, Nov. 2009, pp. 202–205.

[265] T. ISHIKAWA, Y. TSUKUI, and M. MATSUNAMI, "A combined method for the global optimization using radial basis function and deterministic approach", *IEEE Transactions on Magnetics*, vol. 35, pp. 1730–1733, May 1999.

[266] H. M. GUTMANN, “A radial basis function method for global optimization”, *Journal of Global Optimization*, vol. 19, no. 3, pp. 201–227, 2001.

[267] Z. ZHOU, Y. S. ONG, P. B. NAIR, A. J. KEANE, and K. Y. LUM, “Combining global and local surrogate models to accelerate evolutionary optimization”, *IEEE Transactions on Systems, Man, and Cybernetics, Part C (Applications and Reviews)*, vol. 37, no. 1, pp. 66–76, Jan. 2007.

[268] T. OKABE, Y. JIN, and B. SENDHOFFM, “A critical survey of performance indices for multi-objective optimisation”, in *Evolutionary Computation, 2003. CEC ’03. The 2003 Congress on*, vol. 2, Dec. 2003, pp. 878–885.

[269] X. CAI, “A multi-objective gp-pso hybrid algorithm for gene regulatory network modeling”, PhD thesis, Kansas State University, 2009.

[270] H. ISHIBUCHI, N. TSUKAMOTO, and Y. NOJIMA, “Evolutionary many-objective optimization: A short review”, in *Evolutionary Computation, 2008. CEC 2008. (IEEE World Congress on Computational Intelligence). IEEE Congress on*, Jun. 2008, pp. 2419–2426.

[271] S. F. ADRA and P. J. FLEMING, “Diversity management in evolutionary many-objective optimization”, *IEEE Transactions on Evolutionary Computation*, vol. 15, no. 2, pp. 183–195, Apr. 2011.